Lecture Notes in Mathematics

Edited by A. Dold and B. Eckmann

1023

T0220041

Stephen McAdam

Asymptotic Prime Divisors

Springer-Verlag
Berlin Heidelberg New York Tokyo 1983

Author

Stephen McAdam
Department of Mathematics, University of Texas at Austin
Austin, Texas 78712, USA

AMS Subject Classifications (1980): 13A17, 13E05

ISBN 3-540-12722-4 Springer-Verlag Berlin Heidelberg New York Tokyo
ISBN 0-387-12722-4 Springer-Verlag New York Heidelberg Berlin Tokyo

Printing and binding: Beltz Offsetdruck, Hemsbach/Bergstr.
2146/3140-543210

TO MARTHA

ACKNOWLEDGMENTS

Numerous people have participated in the study of asymptotic prime divisors, and I have tried to acknowledge, in the text, a sampling of their contributions. To do so entirely would be impossible, and I hope I have been fair in my selection. Certain people have been particularly helpful to me, as much in stimulating conversations as in specific results. I offer my gratitude to Paul Eakin, Ray Heitmann, Dan Katz and Keith Whittington. My special thanks goes, as it does so often, to Jack Ratliff.

Part of my research was supported by the National Science Foundation, for which I am grateful.

Nita Goldrick typed the manuscript. Her great skill and patience eased a difficult task.

TABLE OF CONTENTS

INTRODUCTION

Asymptotic prime divisors represent the interface of two major ideas in the study of commutative Noetherian rings. The first, the concept of prime divisors, is one of the most valued tools in the researcher's arsenal. The second is the fact that in a Noetherian ring, large powers of an ideal are well behaved, as shown by the Artin-Rees Lemma or the Hilbert polynomial.

Although its roots go back further, the recent interest in asymptotic prime divisors began with a question of Ratliff: What happens to $\operatorname{Ass}(R/I^n)$ as n gets large? He was able to answer a related question, showing that if $\overline{I}$ is the integral closure of I, then $\operatorname{Ass}(R/\overline{I^n})$ stabilizes for large n. In a later work, he also showed that $\operatorname{Ass}(R/\overline{I^n}) \subseteq \operatorname{Ass}(R/\overline{I^{n+1}})$. (Earlier, Rees had shown that if $P \in \operatorname{Ass}(R/\overline{I^n})$, some n, then $P \in \operatorname{Ass}(R/\overline{I^m})$ for infinitely many m.) Meanwhile, Brodmann answered the original question, proving that $\operatorname{Ass}(R/I^n)$ also stabilizes for large n. Since then, the topic of asymptotic prime divisors has been growing rapidly, the latest development being the advent of asymptotic sequences, a useful and interesting analogue of R-sequences.

These notes attempt to present the bulk of the present knowledge of asymptotic prime divisors in a reasonably efficient way, to ease the task of those wishing to learn of, or contribute to the subject. Modulo some gnashing of teeth, and rending of garments, it was both educational and satisfying to write them. I hope that reading them is the same.

The first chapter shows that for an ideal I in a Noetherian ring R, $\operatorname{Ass}(R/I^n)$ stabilizes for large n, as does $\operatorname{Ass}(I^{n-1}/I^n)$, the respective stable values of these two sequences are being denoted $A^*(I)$ and $B^*(I)$. Also $B^*(I)$ is characterized as the contraction to R of prime divisors Q of $t^{-1}\mathcal{R}$ with $It \not\subseteq Q$, where $\mathcal{R} = \mathcal{R}[t^{-1}, It]$ is the Rees ring of R with respect to I.

Chapter Two shows that $A^*(I) - B^*(I) \subseteq \operatorname{Ass} R$, and that $P \in A^*(I) - B^*(I)$ if and only if there is a $k \geq 1$ such that $P^{(k)}$ is part of a primary decomposition of I^n for all sufficiently large n.

Chapter Three shows that $\mathrm{Ass}(R/\overline{I}) \subseteq \mathrm{Ass}(R/\overline{I^2}) \subseteq \ldots$, and that this sequence eventually stabilizes to a set denoted $\bar{A}^*(I)$. Furthermore, $\bar{A}^*(I) \subseteq A^*(I)$. It also developes several technical results useful for dealing with $\bar{A}^*(I)$, the most important of these being that in a local ring, $P \in \bar{A}^*(I)$ if and only if there are primes $q^* \subseteq p^*$ in the completion R^* such that q^* is minimal, $p^* \cap R = P$ and $p^*/q^* \in \bar{A}^*(IR^* + q^*/q^*)$.

In Chapter Four, it is shown that if R is locally quasi-unmixed, then $P \in \bar{A}^*(I)$ if and only if height $P = \ell(I_P)$, the analytic spread of I_P. Since a complete local domain is locally quasi-unmixed, this result meshes nicely with the one mentioned from Chapter Three.

Chapter Five introduces asymptotic sequences: A sequence $x_1, \ldots, x_n$ such that $(x_1, \ldots, x_n) \neq R$ and for $i = 0, \ldots, n-1$, $x_{i+1} \notin \cup \{P \in \bar{A}^*((x_1, \ldots, x_i))\}$. In a local ring (R,M) it is shown that $x_1, \ldots, x_n$ is an asymptotic sequence if and only if $\mathrm{height}((x_1, \ldots, x_n)R^* + q^*/q^*) = n$ for each minimal prime q^* of the completion. This is then used to show that for a given ideal I in any Noetherian ring, all asymptotic sequences maximal with respect to coming from I have the same length, denoted $\mathrm{gr}^* I$. It is then shown that asymptotic sequences are to locally quasi-unmixed rings as R-sequences are to Cohen-Macaulay rings.

In Chapter Six, the sequence $x_1, \ldots, x_n$ is called an asymptotic sequence over the ideal I if $(I, x_1, \ldots, x_n) \neq R$ and for $i = 0, \ldots, n-1$, $x_{i+1} \notin \cup \{P \in \bar{A}^*((I, x_1, \ldots, x_i))\}$. It is shown that in a local ring, all maximal asymptotic sequences over I have the same length.

Chapter Seven proves that in a local ring, the grade of R/I^n stabilizes for large n, and gives partial results concerning $\mathrm{gr}(R/\overline{I^n})$.

Chapter Eight identifies, with one possible exception, all Noetherian rings for which $A^*(I) = \bar{A}^*(I)$ for all ideals I.

In Chapter Nine, asymptotic prime divisors play a minor role in proving the following unexpected result. Let P be prime in a Noetherian domain. Then there is a chain of ideals $P = I_0 \subset I_1 \subset \ldots \subset I_n$ with the following property: Let Q be a prime containing P, and let j be the largest subscript such that $I_j \subseteq Q$. Then $P \subseteq Q$ satisfies going down if and only if j is even.

In Chapter Ten, we consider a local ring (R,M) and the ideal transform of M, $T(M)$. Previously it was known that the following two statements are equivalent: (a) $T(M)$ is an infinite R-module (b) The completion of R contains a depth 1 prime divisor of zero. Our main result adds two more equivalent conditions: (c) $M \in A^*(J)$ for every regular ideal J (d) There is a regular element x with $M \in A^*(J)$ for all $J \sim xR$. Here $J \sim I$ if for some n and m, I^n and J^m have the same integral closure. Motivated by statement (d), we then discuss the possibility of defining a strong asymptotic sequence $x_1, \ldots, x_n$ with $(x_1, \ldots, x_n) \neq R$ and for $i = 0, \ldots, n-1$ $x_{i+1} \notin \cup\{P \in \cap A^*(J) \mid J \sim (x_1, \ldots, x_i)\}$, in the hope that such a sequence will stand in relation to prime divisors of zero, as asymptotic sequences stand to minimal primes. This program is carried out for $n = 1$ and 2.

Chapter Eleven is aptly titled Miscellaneous. It contains topics (of varying worth) which did not fit elsewhere.

The study of asymptotic prime divisors frequently impinges on that of the structure of the spectrum of a Noetherian ring, often referred to as the study of chain conditions. I have tried to keep to a minimum the amount of knowledge of chain conditions necessary to read these notes. In the Appendix, I list those definitions and basic results (with references for the curious reader) which are referred to in the text.

CHAPTER I: $A^*(I)$ and $B^*(I)$

DEFINITION. Let I be an ideal in a Noetherian ring R. For $n = 1,2,3,\ldots,$ let $A(I,n) = \mathrm{Ass}(R/I^n)$ and let $B(I,n) = \mathrm{Ass}(I^{n-1}/I^n)$.

In [R3], Ratliff asked about the behavior of the sequence $A(I,n)$ (and showed that a related sequence stabilized, see Chapter 3). In [B1], Brodmann showed that both sequences $A(I,n)$ and $B(I,n)$ stabilize for large n, as we now show. Recall that the graded Noetherian ring $T = \Sigma R_n$, $n \geq 0$ is homogeneous if $T = R_0[R_1]$. Our first lemma is well known.

LEMMA 1.1. a) Let $\Sigma_{n\geq 0} R_n$ be a Noetherian homogeneous graded ring. Then there is an ℓ such that for $n \geq \ell$, $(0 : R_1) \cap R_n = 0$.

b) Let I be an ideal in a Noetherian ring. Then there is an ℓ such that for $n \geq \ell$, $(I^{n+1} : I) \cap I^{\ell} = I^n$.

Proof: a) Let $(0 : R_1)$ be generated by homogeneous elements $a_1, \ldots, a_s$. Let $\ell = 1 + \max\{\deg a_i\}$. If $x = \Sigma r_i a_i \in (0 : R_1) \cap R_n$ with $n \geq \ell$, then we may assume the r_i are homogeneous and have $r_i \in R_1 T$. Thus $r_i a_i = 0$, and so $x = 0$.

b) Let $\Sigma R_n = \Sigma I^n/I^{n+1}$ and pick ℓ as above. Say $n \geq \ell$ and let $x \in (I^{n+1} : I) \cap I^{\ell}$. Suppose that $x \notin I^n$. Let $x \in I^k - I^{k+1}$, and note that $\ell \leq k < n$. Since $xI \subseteq I^{n+1} \subseteq I^{k+2}$, with $\bar{x} \in I^k/I^{k+1} = R_k$ we have $0 \neq \bar{x} \in (0 : R_1) \cap R_k$, contradicting part a.

LEMMA 1.2. Let $T = \Sigma R_n$ be a Noetherian graded ring. Let I be a homogeneous ideal and let c be a homogeneous element. Suppose that S is a multiplicatively closed subset of R_0 and that $(I : c) \cap S = \emptyset$. Then there is a homogeneous element d, such that $(I : cd)$ is prime and $(I : cd) \cap S = \emptyset$.

Proof: Among all homogeneous d' with $(I : cd') \cap S = \emptyset$, choose d so that $(I : cd)$ is maximal. It is enough to take homogeneous x and y not in $(I : cd)$ and show $xy \notin (I : cd)$. Suppose, contrarily, that $xy \in (I : cd)$. Then $x \in (I : cdy)$ so that $(I : cdy)$ is strictly larger than $(I : cd)$. Thus there is an $s \in S \cap (I : cdy)$. Now $y \in (I : cds)$, showing that this ideal is strictly larger than $(I : cd)$. Thus there is an $s' \in S \cap (I : cds)$. This gives $ss' \in S \cap (I : cd)$, a contradiction.

PROPOSITION 1.3. [ME] Let $T = \Sigma_{n \geq 0} R_n$ be a Noetherian homogeneous graded ring. Then there exists an m such that $\mathrm{Ass}_{R_0}(R_m) = \mathrm{Ass}_{R_0}(R_n)$ for all $n \geq m$.

Proof: Let $P \in \bigcup \mathrm{Ass}_{R_0}(R_k)$, $k = 0,1,2,\ldots$. Then for some homogeneous $c \in T$, $P = (0 : c)_{R_0}$. Clearly $P = (0 : c)_T \cap R_0$ and by Lemma 1.2, for some homogeneous $d \in T$ we have $P^* = (0 : cd)$ prime in T and $P^* \cap R_0 = P$. As $\mathrm{Ass}_T(T)$ is finite, we see that $\bigcup \mathrm{Ass}_{R_0}(R_k)$ is finite.

Now select ℓ as in Lemma 1.1 and say $n \geq \ell$. If $P \in \mathrm{Ass}_{R_0}(R_n)$ write $P = (0 : c)_{R_0}$, $c \in R_n$. As $n \geq \ell$, $P = (0 : cR_1)_{R_0}$. Since $cR_1 \subseteq R_{n+1}$, we have $P \in \mathrm{Ass}_{R_0}(R_{n+1})$. Thus $\mathrm{Ass}_{R_0}(R_n) \subseteq \mathrm{Ass}_{R_0}(R_{n+1})$ for $n \geq \ell$. As we already have $\bigcup \mathrm{Ass}_{R_0}(R_k)$ finite, the result follows.

COROLLARY 1.4. (Brodmann [B1]) Let I be an ideal in the Noetherian ring R. The sequence $B(I,n)$ stabilizes.

Proof: Apply the proposition to $\Sigma I^{n-1}/I^n$.

COROLLARY 1.5. (Brodmann [B1]) Let I be an ideal in the Noetherian ring R. The sequence $A(I,n)$ stabilizes.

Proof: The exact sequence $0 \to I^n/I^{n+1} \subseteq R/I^{n+1} \to R/I^n \to 0$ shows that $A(I,n+1) \subseteq A(I,n) \cup B(I,n+1)$. For large n, we already have $B(I,n+1) = B(I,n) \subseteq A(I,n)$. Thus $A(I,n+1) \subseteq A(I,n)$, and the result is clear since $A(I,n)$ is finite.

Note that for an ideal I in a Noetherian ring R, $B(I,n) \subseteq A(I,n)$. The following example, due to A. Sathaye, shows that neither sequence is monotone.

EXAMPLE. Let k be a field and n a positive integer. Let $R = k[x, z_1, \ldots, z_{2n}]$ with the restrictions that $xz_{2i-1}^{2i-1} = z_{2i}^{2i}$ for $i = 1,2,\ldots,n$, and $z_j^j z_i = 0$ for

$1 \leq i,j \leq 2n$. Let $I = (z_1, z_2, \ldots, z_{2n}) \subseteq P = (x, z_1, \ldots, z_{2n})$. Then for $1 \leq i \leq 2n$, $P \in B(I,i)$ if i is even, while $P \notin A(I,i)$ if i is odd.

Proof: Since $z_{2i-1}^{2i-1} \notin I^{2i}$ and $Pz_{2i-1}^{2i-1} \subseteq I^{2i}$, we have $P \in B(I,2i)$ for $1 \leq i \leq n$. To see that $P \notin A(I,s)$ for s odd, $1 \leq s \leq 2n$, note that $P \notin A(I,1)$ since I is prime. Now for $1 \leq q \leq 2n$, the residues of the set $T_q = \{z_q^q\} \cup \{z_2^{u_2} \ldots z_{2n}^{u_{2n}} | u_2 + \ldots + u_{2n} = q,\ 0 \leq u_i < i\}$ form a generating set for I^q/I^{q+1} over $k[x]$. If q is even, there are no relations, and T_q gives a free basis. If q is odd, there is the unique relation $xz_q^q \in I^{q+1}$. Suppose $P \in A(I,s)$. Then for some $w \notin I^s$, $Pw \subseteq I^s$. Consider r such that $w \in I^r - I^{r+1}$. By the previous remarks, $xw \in I^{r+1}$ shows that r is odd. Furthermore, it can be seen that $xw \notin I^{r+2}$. Thus $s = r+1$ and so s is even.

DEFINITION. For I an ideal in a Noetherian ring, the eventual constant values of the sequences $A(I,n)$ and $B(I,n)$ will be denoted $A^*(I)$ and $B^*(I)$, respectively.

The fact that $A^*(I)$ and $B^*(I)$ behave well under localization is straightforward, and yet we will use it so often that we state it formally.

LEMMA 1.6. Let $I \subseteq P$ be ideals in a Noetherian ring, with P prime. Then $P \in A^*(I)$ (respectively $P \in B^*(I)$) if and only if $P_S \in A^*(I_S)$ (respectively $P_S \in B^*(I_S)$), for any multiplicatively closed set S disjoint from P.

The next result will lead to some interesting applications of asymptotic prime divisors. As this result will be used again when discussing the integral closure of an ideal (Chapter 3), we give it here in full generality.

If J is an ideal of R, we will use $\bar{J}$ to denote the integral closure of J. Thus $\bar{J} = \{x \in R | x$ satisfies a polynomial of form $X^n + j_1X^{n-1} + \ldots + j_n = 0$, with $j_i \in J^i\}$. Recall that $\bar{R}$ is the integral closure of R.

PROPOSITION 1.7. [M3] Let P be a prime ideal in a Noetherian domain R. There is an integer $n \geq 1$ with the following property: If I is an ideal of R with

$I \subseteq \overline{P^n}$, and if there exists an integral extension domain T of R and a $Q \in \operatorname{spec} T$ with $Q \cap R = P$ and Q minimal over IT, then $P \in \operatorname{Ass}(R/I)$.

Proof: Let $P_1, \ldots, P_m$ be all of the primes of $\overline{R}$ which lie over P. Select $u_i \in P_i - \cup P_j$, $j \neq i$, and let $S = R[u_1, \ldots, u_m]$. Notice that P_i is the unique prime of $\overline{R}$ lying over $p_i = P_i \cap S$. Let (V_i, N_i) be a D.V.R. overring of S with $N_i \cap S = p_i$. Since S is a finitely generated R-module, we can choose $b \in R$ with $bS \subseteq R$. Pick n sufficiently large that $b \notin N_i^n$, $i = 1,2,\ldots,m$.

Suppose that $I \subseteq \overline{P^n}$ and that T is an integral extension domain of R containing a prime Q with $Q \cap R = P$ and Q minimal over IT. We first reduce to the case that $T = S$. Clearly we may assume $T = \overline{T}$, and by going down we may replace $\overline{T}$ by $\overline{R}$. Finally since P_i is the only prime of $\overline{R}$ lying over p_i, $i = 1,2,\ldots,m$, by going up we replace $\overline{R}$ by S.

We now have $T = S$, and of course $Q = p_i$ for some $i = 1,2,\ldots,m$. We localize making P maximal in R. Since p_i is minimal over IS, there is an integer $k \geq 1$ and an $s \in S - p_i$ with $sp_i^k \subseteq IT$. Using $bS \subseteq R$, we have $bsP^k \subseteq bsp_i^k \subseteq bIT \subseteq I$. Furthermore, we claim $bs \notin \overline{I}$. If $bs \in \overline{I} \subseteq \overline{P^n} \subseteq \overline{p_i^n} \subseteq \overline{N_i^n} = N_i^n$, then since $s \notin p_i$ implies s is a unit of V_i, we have $b \in N_i^n$, contradicting our choice of n. Thus $bs \notin I$ but $bsP^k \subseteq I$, showing that P^k consists of zero divisors modulo I. As P was maximal, $P \in \operatorname{Ass}(R/I)$.

COROLLARY 1.8. Let I be an ideal in a Noetherian domain R and let T be an integral extension domain of R. If Q is prime in T and minimal over IT, then $Q \cap R \in A^*(I)$.

Proof: Let $P = Q \cap R$, and choose n as in the proposition. Then $P \in A(I,m)$ for $m \geq n$.

The following fact about the integral closure of a Noetherian domain appears to depend upon knowledge of asymptotic prime divisors.

PROPOSITION 1.9. Let R be a Noetherian domain. Let $J = (b_1, \ldots, b_m)$ be a finitely generated ideal of the integral closure $\bar{R}$. Then the number of primes of $\bar{R}$ minimal over J is finite.

Proof: Let $S = R[b_1, \ldots, b_m]$ and let $I = (b_1, \ldots, b_m)S$. Thus S is Noetherian and $I\bar{R} = J$. If $Q \in \operatorname{spec} \bar{R}$ and Q is minimal over J, then by Corollary 1.8, $Q \cap S \in \bar{A}^*(I)$. Since $\bar{A}^*(I)$ is finite and since only finitely many primes of $\bar{R}$ lie over a given prime on S, we are done.

We generalize [N, 33.11].

PROPOSITION 1.10. Let $R \subseteq T$ be an integral extension of domains with R Noetherian. Let Q be a height n prime of T and let $P = Q \cap R$. Then grade $P \leq n$. If grade $P = n$, then for any R-sequence $a_1, \ldots, a_n$ coming from P, P is a prime divisor of $(a_1, \ldots, a_n)$.

Proof: We induct on n. For $n = 1$, pick $a \neq 0$ in P. Since height $Q = 1$, Q is minimal over aT. By Corollary 1.8, for sufficiently large k, P is a prime divisor of $a^k R$. It is not difficult to now see that P is also a prime divisor of aR.

For $n > 1$, suppose grade $P \geq n$ and let $a_1, \ldots, a_n$ be an R-sequence coming from P. We claim height$(a_1, \ldots, a_n)T = n$. If not, say $q \in \operatorname{spec} T$, height $q < n$ and $(a_1, \ldots, a_n)T \subseteq q$. By induction, grade $q \cap R \leq$ height $q < n$, contradicting that $a_1, \ldots, a_n$ is an R-sequence in $q \cap R$. Thus the claim is true, and so Q is minimal over $(a_1, \ldots, a_n)T$. By Corollary 1.8, for large k we have P a prime divisor of $(a_1, \ldots, a_n)^k$ in R. As $a_1, \ldots, a_n$ is an R-sequence, P is also a prime divisor of $(a_1, \ldots, a_n)$ by [K1, Section 3-1, Exercise 13].

In Chapter 5 we strengthen Proposition 1.10, replacing "height Q" by "little height Q".

The next three propositions give easy circumstances under which a prime must be in $\bar{A}^*(I)$.

PROPOSITION 1.11. Let I be an ideal in a Noetherian ring, and let the prime P be minimal over I. Then $P \in A^*(I)$. Also $P \in B^*(I)$ if and only if height $P > 0$.

Proof: Since P is minimal over I^n for all n, $P \in A(I,n)$, and so $P \in A^*(I)$. For the second statement, localize at P, so that I is P-primary. Now height $P > 0$ if and only if I is not nilpotent. If I is nilpotent, clearly $P \notin B^*(I)$. If I is not nilpotent, then for all n I^n/I^{n+1} is a nonzero module (by Nakayama's Lemma) which must have at least one prime divisor. However P is the only possibility. Thus $P \in B(I,n)$ for all n.

PROPOSITION 1.12. Let $I \subseteq P$ with P a prime divisor of 0 in a Noetherian ring. Then $P \in A^*(I)$.

Proof: Localize at P and then write $P = (0 : c)$. For n large enough that $c \notin I^n$, clearly $P = (I^n : c)$.

Our next proposition generalizes Proposition 1.11. The lemma is due to Ratliff.

LEMMA 1.13. Let $Q \subset P$ be primes of the Noetherian ring R such that Q is a prime divisor of 0. Then there is an integer $n > 0$ such that for any ideal J of R with $J \subseteq P^n$ and P minimal over $Q + J$, we have $P \in \mathrm{Ass}(R/J)$.

Proof: Localize at P. Let $q_1 \cap \ldots \cap q_r$ be a primary decomposition of 0 with q_1 primary to Q. Choose $0 \neq x \in q_2 \cap \ldots \cap q_r$, and pick n such that $x \notin P^n$. Suppose that $p \in \mathrm{Ass}(R/J)$ and $p \neq P$. Since P is minimal over $Q + J$, we have $Q \not\subseteq p$. Thus in R_p, $0 = (q_2)_p \cap \ldots \cap (q_r)_p$ so that $xR_p = 0$. This shows that x is in every p-primary ideal. However, $J \subseteq P^n$ shows that $x \notin J$. Thus $P \in \mathrm{Ass}(R/J)$, using primary decomposition.

PROPOSITION 1.14. Let I, P, Q be ideals in a Noetherian ring with Q a prime divisor of 0, and P a prime minimal over $Q + I$. Then $P \in A^*(I)$.

Proof: With n as in Lemma 1.13, $P \in A(I,m)$ for all $m \geq n$.

Later (Proposition 2.5) we will strengthen Proposition 1.14 to say that if in addition $P \neq Q$, then $P \in B^*(I)$.

We give a characterization of $B^*(I)$ in terms of the Rees ring of R with respect to I, that is, the ring $\mathcal{R} = R[t^{-1}, It]$ with t an indeterminate.

PROPOSITION 1.15. Let I be an ideal in the Noetherian ring R, and let $\mathcal{R} = R[t^{-1}, It]$ be the Rees ring of R with respect to I. Then $P \in B^*(I)$ if and only if there is a prime divisor Q of $t^{-1}\mathcal{R}$ such that $It \not\subseteq Q$ and $Q \cap R = P$.

Proof: Let $P \in B^*(I)$. Consider ℓ as in Lemma 1.1b, and choose $n > \ell$ with $P \in \mathrm{Ass}(I^n/I^{n+1})$. Write $P = (I^{n+1} : c)$ with $c \in I^n$. Since $ct^n \in \mathcal{R}$, note that $(t^{-1}\mathcal{R} : ct^n) \cap R = (I^{n+1} : c) = P$. By Lemma 1.2 there is a $dt^m \in \mathcal{R}$ such that $Q = (t^{-1}\mathcal{R} : dct^{n+m})$ is prime in $\mathcal{R}$ and $Q \cap R = P$. We must show that $It \not\subseteq Q$. Since Q is a proper ideal, $dct^{n+m} = (dt^m)(ct^n) \notin t^{-1}\mathcal{R}$. Thus $m \geq 0$ and $dc \notin I^{n+m+1}$. By Lemma 1.1, $(I^{n+m+2} : I) \cap I^{\ell} = I^{n+m+1}$, and since $c \in I^n$ we must have $dc \notin (I^{n+m+2} : I)$. Therefore $It \not\subseteq (t^{-1}\mathcal{R} : dct^{n+m}) = Q$ as desired.

Conversely, suppose that $Q = (t^{-1}\mathcal{R} : gt^k)$ with $g \in I^k$, that $Q \cap R = P$, and that $It \not\subseteq Q$. Pick $ht \in It - Q$. Clearly $Q = (t^{-1}\mathcal{R} : gh^m t^{k+m})$ for all $m > 0$. Thus $P = (I^{k+m+1} : gh^m)$. Since $gh^m \in I^{k+m}$ and m is arbitrary, we have $P \in B^*(I)$.

We close the chapter with a question. We have seen that the sequence $A(I,1), A(I,2), \ldots$ is not increasing. Ratliff asks whether $A(I,1) \cap A^*(I)$, $A(I,2) \cap A^*(I), \ldots$ is increasing?

CHAPTER II: $A^*(I) - B^*(I)$

In this chapter, we study primes contained in $A^*(I)$ but not $B^*(I)$, our main result being that such primes must be prime divisors of zero.

LEMMA 2.1. Let I be an ideal in a Noetherian ring R and (by Lemma 1.1) suppose for $n > \ell$ we have $(I^n : I) \cap I^\ell = I^{n-1}$. If P is prime in R and if $P = (I^n : c)$ with $n > \ell$ and $c \in I^\ell$, then $P \in B^*(I)$.

Proof: Since $cI \subseteq cP \subseteq I^n$, we have $c \in (I^n : I) \cap I^\ell = I^{n-1}$. For $j \geq 0$, clearly $P \subseteq (I^{n+j} : cI^j)$. Conversely, if $r \in (I^{n+j} : cI^j)$ then $rcI^{j-1} \subseteq (I^{n+j} : I) \cap I^\ell = I^{n+j-1}$, so that $r \in (I^{n+j-1} : cI^{j-1})$. Iterating, we find $r \in (I^n : c) = P$. Thus $P = (I^{n+j} : cI^j)$ for $j = 1, 2, \ldots$. Now we already have $c \in I^{n-1}$, so $cI^j \subseteq I^{n+j-1}$. Thus $P \in B(I, n+j)$ for $j = 1, 2, \ldots$.

PROPOSITION 2.2 [ME] Let I be an ideal in a Noetherian ring R. If $P \in A^*(I) - B^*(I)$ then P is a prime divisor of zero.

Proof: We may localize at P. Since $P \in A^*(I)$, for all large n we have an $x_n \in R$ with $P = (I^n : x_n)$, and by Lemma 2.1 we have $x_n \notin I^\ell$. To show that P is a prime divisor of zero, it is sufficient to show this in the case that (R,P) is complete, which we now assume. Let $V = (I^\ell : P)/I^\ell$ and for $n > \ell$ let $V_n = [(I^n : P) + I^\ell]/I^\ell$. Now $PV = 0$, so V is a finite dimensional vector space over R/P. Clearly x_n taken modulo I^ℓ is a nonzero element in the subspace V_n. Since $V_{n+1} \subseteq V_n$, we see that $\cap V_n \neq 0$, by finite dimensionality. Let $\bar{\lambda} \neq 0$ be in this intersection, and let $\lambda \in (I^\ell : P) - I^\ell$ be a preimage. Since $\bar{\lambda} \in V_n$ write $\lambda = d_n + i_n$ with $d_n \in (I^n : P)$ and $i_n \in I^\ell$. For $m \geq n$ we have $I(i_n - i_m) \subseteq P(i_n - i_m) = P(d_n - d_m) \subseteq I^n$. Thus $i_n - i_m \in (I^n : I) \cap I^\ell = I^{n-1} \subseteq P^{n-1}$, showing that the sequences $\{i_n\}$ and $\{d_n\}$ are Cauchy sequences. Let $i_n \to i$ and $d_n \to d$. Since $i \in I^\ell$ and $\lambda \notin I^\ell$, $d \neq 0$. Finally, since $d_n P \subseteq I^n \subseteq P^n$, $dP \subseteq \cap P^n = 0$, concluding the proof.

PROPOSITION 2.3. Let I be an ideal in a Noetherian ring R. Then $P \in A^*(I) - B^*(I)$ if and only if there is an integer $k \geq 1$ such that for all sufficiently large n, the k-th symbolic power $P^{(k)}$ is part of a primary decomposition of I^n.

Proof: We may assume R is local at P. In this context, $P^{(k)} = P^k$. First suppose that for all large n, that I^n has primary decomposition $I^n = P^k \cap q_{n1} \cap \ldots \cap q_{nm}$. Clearly $P \in A^*(I)$. Let $n \geq k+1$ and let $P = (I^n : c)$. To show $P \notin B^*(I)$, we desire $c \notin I^{n-1}$. Since q_{ni} is primary to a prime properly contained in P, $Pc \subseteq I^n$ shows that $c \in q_{n1} \cap \ldots \cap q_{nm}$. Since $P = (I^n : c)$ clearly $c \notin I^n$. Thus $c \notin P^k$. However $I^{n-1} \subseteq P^k$, so $c \notin I^{n-1}$ as desired.

Conversely, suppose that $P \in A^*(I) - B^*(I)$. Let ℓ be as in Lemma 1.1. Since ℓ can be increased, also assume that $A(I,\ell) = A(I,\ell+1) = \ldots = A^*(I)$. Suppose $A^*(I) = \{P, Q_1, \ldots, Q_m\}$ and for $n \geq \ell$ let the primary decomposition of I^n be $q_n \cap q_{n1} \cap \ldots \cap q_{nm}$ with $\mathrm{RAD}\, q_n = P$. We may assume that $q_{n+1i} \subseteq q_{ni}$, so that if $J_n = q_{n1} \cap \ldots \cap q_{nm}$, then $J_{n+1} \subseteq J_n$. Claim $J_n \cap I^\ell = I^n$. One inclusion is obvious. Consider $c \in J_n \cap I^\ell$, and by way of contradiction, assume $c \notin I^n$. Thus $(I^n : c)$ is proper, and so is contained in P. Now for some large t, $P^t \subseteq q_n$. Thus $P^t c \subseteq q_n c \subseteq q_n \cap J_n = I^n$. Therefore $P^t \subseteq (I^n : c) \subseteq P$. Using Lemma 1.2, we see that for some $r \in R$, $(I^n : rc)$ is a proper prime ideal, and since $P^t \subseteq (I^n : c) \subseteq (I^n : rc)$, we must have $(I^n : rc) = P$. Now $c \in I^\ell$, so $rc \in I^\ell$, and by Lemma 2.1 we have $P \in B^*(I)$. This is a contradiction, proving the claim that $J_n \cap I^\ell = I^n$. Now since $J_{n+1} \subseteq J_n$, we have $J_n \subseteq J_\ell$. Thus $I^n = J_n \cap I^\ell = J_n \cap J_\ell \cap q_\ell = J_n \cap q_\ell$. Finally, take k large enough that $P^k \subseteq q_\ell$. Then for $n \geq k$, $I^n = J_n \cap P^k$, giving the desired primary decomposition.

Consider an ideal I in a Noetherian ring R. Since $(0 : I) \subseteq (0 : I^2) \subseteq \ldots$, there is a k with $(0 : I^k) = (0 : I^{k+1}) = \ldots$. Call this ideal J. The following lemma, due to P. Eakin, shows that reducing modulo J can be useful.

LEMMA 2.4. Consider I and $J = (0 : I^k)$ as above. Let $I' = I$ modulo J. Then $A^*(I') = B^*(I')$. Also $P \in B^*(I)$ if and only if $J \subseteq P$ and $P' \in A^*(I')$.

Proof: Suppose $P' \in A^*(I') - B^*(I')$. Then by Proposition 2.2, $P' = (0 : c')$ for some $c' \in R'$. Since $I' \subseteq P'$, we have $c'I' = 0$. Thus $cI \subseteq J = (0 : I^k)$, so that $cI^{k+1} = 0$. Therefore $c \in (0 : I^{k+1}) = (0 : I^k) = J$, and so $c' = 0$. This contradicts that $(0 : c') = P' \neq R'$, and shows that $A^*(I') = B^*(I')$.

Now consider ℓ as in Lemma 1.1. We easily see that $(I^{n+k} : I^k) \cap I^\ell = I^n$ for large n. Suppose that $x \in J \cap I^\ell$. Then $xI^k = 0 \subseteq I^{n+k}$ for all big n, so that $x \in (I^{n+k} : I^k) \cap I^\ell = I^n$ for all big n. That is, $J \cap I^\ell \subseteq \cap I^n$, $n = 1, 2, \ldots$.

We next claim that for large n, $I^n/I^{n+1} \approx I'^n/I'^{n+1}$. To see this, map $a + I^{n+1}$ to $a' + I'^{n+1}$ and note that the map is injective since $I^n \cap J \subseteq I^{n+1}$, from above.

Suppose $P \in B^*(I)$. This is still true upon localizing at P, and so I_P is not nilpotent. Thus $J \subseteq P$, or else we would have $I_P^k = 0$. By the above isomorpism, P is a prime divisor on I'^n/I'^{n+1}, for large n. However, J annihilates this module, which therefore is an R'-module having P' as a prime divisor. Thus $P' \in B^*(I')$. The converse is similar.

We use this lemma to strengthen Proposition 1.14.

PROPOSITION 2.5. Let I, P, Q be ideals in a Noetherian ring R, with Q a prime divisor of zero and with P a prime minimal over $Q + I$. If $P \neq Q$, then $P \in B^*(I)$.

Proof: Let $J = (0 : I^k)$ as in Lemma 2.4. Since $P \neq Q$, clearly $I \not\subseteq Q$, and so $J \subseteq Q$. Writing $Q = (0 : c)$ for some $c \in R$, we have $Q \subseteq (J : c)$. We claim equality. If $x \in (J : c)$ then since $I \not\subseteq Q$, pick $y \in I^k - Q$. As $xc \in J = (0 : I^k)$, we have $xcy = 0$. That is, $xy \in (0 : c) = Q$. Since $y \notin Q$, $x \in Q$ as claimed.

We now work modulo J, using primes to denote images. We have just seen that Q' is a prime divisor or zero in R'. As P' is minimal over $Q' + I'$, Proposition 1.14 gives $P' \in A^*(I')$. By Lemma 2.4, $P \in B^*(I)$.

Combining Propositions 2.2 and 1.12, we see that $A^*(I) = B^*(I) \cup \{P \in \mathrm{Ass}(R) \mid I \subseteq P\}$. We know very little of what can be said about the overlap, $B^*(I) \cap \{P \in \mathrm{Ass}(R) \mid I \subseteq P\}$. We give an example in which $I \subset P_1 \subset P_2 \subset P_3$ with P_1, P_2, P_3 all in $\mathrm{Ass}(R)$, and P_1 and P_3 are in $B^*(I)$, but $P_2 \notin B^*(I)$.

EXAMPLE. Let $T = K[X,Y,Z,W]$. Let $p_1 = (X,Y)$, $p_2 = (X,Y,Z)$, and $p_3 = (X,Y,Z,W)$, while $q_1 = (X)$, $q_2 = (Z)$, and $q_3 = (X,Z,W)$. Let $L = p_1p_2p_3q_1q_2q_3$, let $R = T/L$, let $P_1, P_2, P_3, Q_1, Q_2, Q_3$ be (respectively) the images of $p_1, p_2, p_3, q_1, q_2, q_3$ and let I be the image of (Y). Now $I \subseteq P_1 \subseteq P_2 \subseteq P_3$ and these three primes are in $\mathrm{Ass}(R)$. Also Q_1 and Q_3 are in $\mathrm{Ass}(R)$. Since $P_1 = Q_1 + I$ and $P_3 = Q_3 + I$, by Proposition 2.5 we have P_1 and P_3 in $B^*(I)$. We claim that $P_2 \notin B^*(I)$. Invoking Lemma 2.4, we note that $Y^3 \in p_1p_2p_3$, so for $k \geq 3$ we have $(0 : I^k) = Q_1Q_2Q_3$. By Lemma 2.4, we must show that $p_2/q_1q_2q_3 \notin A^*\left(\frac{(Y)+q_1q_2q_3}{q_1q_2q_3}\right)$. However the image of Y modulo $q_1q_2q_3$ is a regular element, and so we need only show that $p_2/q_1q_2q_3$ is not a prime divisor of $(Y) + q_1q_2q_3/q_1q_2q_3$, which is clear.

It is natural to ask whether in Proposition 2.5, the condition P minimal over $Q + I$ can be weakened to just $P \in B^*(Q + I)$. The answer is no. In the above example, we already have $P_2 \notin B^*(I)$. However, since $P_2 = Q_2 + (Q_1 + I)$, Proposition 2.5 shows that $P_2 \in B^*(Q_1 + I)$.

CHAPTER III: $\overrightarrow{A}^*(I)$

In this chapter, we study another sort of asymptotic prime divisor. Recall that for an ideal I, $\overline{I}$ denotes the integral closure of I. We will let $\overline{A}(I,n) = \operatorname{Ass}(R/\overline{I^n})$. Our main goals will be to prove that $\overline{A}(I,1) \subseteq \overline{A}(I,2) \subseteq \ldots$, that this sequence eventually stabilizes at a set denoted $\overrightarrow{A}^*(I)$, and that $\overrightarrow{A}^*(I) \subseteq A^*(I)$. These results were first proved by Ratliff in [R3] and [R8]. (In [Rs2], Rees outlines a different approach to these ideas.) The essence of the arguments needed are mostly easily seen when R is a domain. We treat that case first, following Ratliff's trail.

LEMMA 3.1. Let R be a Noetherian domain with integral closure $\overline{R}$. Let (V,N) be a D.V.R. overring of R and suppose that V is a localization of an integral extension of a finitely generated extension of R. If the transcendence degree of V/N over $R/N \cap R$ is 0, then $\operatorname{height}(N \cap \overline{R}) = 1$.

Proof: Let V be a localization of an integral extension of A, with A finitely generated over R. Now $A \subseteq \overline{A} \subseteq V$, and $\overline{A}$ is an integral extension of a finitely generated extension of $\overline{R}$. Since the transcendence degree of $\overline{A}/N \cap \overline{A}$ over $\overline{R}/N \cap \overline{R}$ is 0, and also since $\operatorname{height}(N \cap \overline{A}) = \operatorname{height} N = 1$, we see that $N \cap \overline{A}$ is isolated among primes of $\overline{A}$ lying over $N \cap \overline{R}$. By the Peskine-Evans formulation of Zariski's Main Theorem [E], $\overline{R}_{N \cap \overline{R}} = \overline{A}_{N \cap \overline{A}}$. Thus $\operatorname{height}(N \cap \overline{R}) = \operatorname{height}(N \cap \overline{A}) = 1$.

LEMMA 3.2. Let $a \neq 0$ in the Noetherian domain R. Let P be a prime divisor of $\overline{a^n R}$ for some $n > 0$. Then in $\overline{R}$ there is a height 1 prime divisor p of $a\overline{R}$ with $p \cap R = P$.

Proof: Write $P = (\overline{a^n R} : c)$. Since $\overline{a^n R} = a^n\overline{R} \cap R$, $P = (a^n\overline{R} : c)_{\overline{R}} \cap R$. Now the Krull domain $\overline{R}$ has A.C.C. on ideals of the form $(b : d)$. Thus the argument used in proving Lemma 1.2 shows that $(a^n\overline{R} : c)_{\overline{R}}$ can be enlarged to a prime divisor p of $a^n\overline{R}$, with $p \cap R = P$. Since $\overline{R}$ is a Krull domain, $\operatorname{height} p = 1$.

LEMMA 3.3. Let I be an ideal in a Noetherian domain R. Let $\mathcal{R} = R[t^{-1}, It]$ be the Rees ring of R with respect to I. If Q is a prime divisor of $\overline{t^{-n}\mathcal{R}}$ for some $n > 0$, then either $It \not\subseteq Q$ or $Q \cap R = 0$.

Note: Since $I = t^{-1}\mathcal{R} \cap R \subseteq Q \cap R$, if $I \neq 0$ we must have $It \not\subseteq Q$.

Proof: Using Lemma 3.2 we see that $\overline{\mathcal{R}}$ contains a height 1 prime q lying over Q. Now $(V,N) = (\overline{\mathcal{R}}_q, q_q)$ is a D.V.R. Suppose that $It \subseteq Q$. Then $R/Q \cap R = \mathcal{R}/Q$ and clearly the transcendence degree of V/N over $R[t^{-1}]/Q \cap R[t^{-1}]$ is 0. By Lemma 3.1 we have height $N \cap \overline{R}[t^{-1}] =$ height $Q \cap \overline{R}[t^{-1}] = 1$. Moreover $Q \cap R[t^{-1}]$ obviously equals $(Q \cap R, t^{-1})R[t^{-1}]$, since $t^{-1} \in Q$. However, for T a Noetherian domain and X an indeterminate, only one height 1 prime of $\overline{T}[X]$ contains X, namely $X\overline{T}[X]$, and it intersects $T[X]$ at $XT[X]$. With $T = R$ and $X = t^{-1}$, we see that $(Q \cap R, t^{-1})R[t^{-1}]$ must be just $t^{-1}R[t^{-1}]$, so that $Q \cap R = 0$.

PROPOSITION 3.4. Let $I \neq 0$ be an ideal in a Noetherian domain R. Then $\overline{A}(I,1) \subseteq \overline{A}(I,2) \subseteq \dots$. This sequence stabilizes to a set denoted $\overline{A}^*(I)$. Furthermore $\overline{A}^*(I) \subseteq B^*(I)$.

Proof: Let $P \in \overline{A}(I,n)$ and write $P = (\overline{I^n} : c)_R$ for $c \in R$. Then since $\overline{I^n} = \overline{t^{-n}\mathcal{R}} \cap R$, $P = (\overline{t^{-n}\mathcal{R}} : c)_{\mathcal{R}} \cap R$. By Lemma 3.3, no prime divisor of $(\overline{t^{-n}\mathcal{R}} : c)_{\mathcal{R}}$ can contain It. Select $dt \in It$ with dt a nonzero divisor modulo $(\overline{t^{-n}\mathcal{R}} : c)_{\mathcal{R}}$. Thus $(\overline{t^{-n}\mathcal{R}} : c)_{\mathcal{R}} = (\overline{t^{-n}\mathcal{R}} : cdt)_{\mathcal{R}}$ and this ideal meets R at P. Since the degree 1 component of $\overline{t^{-n}\mathcal{R}}$ is $(\overline{I^{n+1}} \cap I)t$, we see that $P = (\overline{I^{n+1}} \cap I : cd) = (\overline{I^{n+1}} : cd)$ using that $d \in I$. Therefore $P \in \overline{A}(I,n+1)$, which shows the sequence is increasing.

We will now show that $\overline{A}(I,n) \subseteq B^*(I)$. As $B^*(I)$ is finite, the rest of the theorem follows immediately. As above, write $P = (\overline{t^{-n}\mathcal{R}} : c)_{\mathcal{R}} \cap R$. As in Lemma 1.2, enlarge $(\overline{t^{-n}\mathcal{R}} : c)_{\mathcal{R}}$ to a prime divisor Q of $\overline{t^{-n}\mathcal{R}}$ with $Q \cap R = P$. By Lemma 3.3, $It \not\subseteq Q$. Also Lemma 3.2 gives a height 1 prime q of $\overline{\mathcal{R}}$ lying over Q. By Proposition 1.10, height $q = 1$ implies grade $Q = 1$. Thus Q is a prime divisor of $t^{-1}\mathcal{R}$. Since $It \not\subseteq Q$, Proposition 1.15 tells us that $P = Q \cap R \in B^*(I)$.

We apply Proposition 1.7 to this context, and easily see the following.

PROPOSITION 3.5. Let $R \subset T$ be an integral extension of domains, with R Noetherian. Let I be an ideal of R. If $Q \in \operatorname{spec} T$ with Q minimal over It, then $Q \cap R \in \overrightarrow{A}^*(I)$.

We now drop the assumption that R is a domain.

LEMMA 3.6. If $I \subset J$ are ideals, then I reduces J if and only if for each minimal prime q, $I+q/q$ reduces $J+q/q$.

Proof: One direction is trivial. Thus suppose that $q_1, \ldots, q_m$ are all the minimal primes, and that for each i, $I \bmod q_i$ reduces $J \bmod q_i$. Then for sufficiently large k we have $J^{k+1} \subseteq IJ^k + q_i$, $i = 1, \ldots, m$. Since $IJ^k \subseteq J^{k+1}$, in fact we have $J^{k+1} = IJ^k + (J^{k+1} \cap q_i)$. If n is such that $(q_1 \cdots q_m)^n = 0$, then $J^{nm(k+1)} = \prod_{i=1}^{m} (IJ^k + (J^{k+1} \cap q_i))^n \subseteq IJ^{nm(k+1)-1} \subseteq J^{nm(k+1)}$. Thus I reduces J.

LEMMA 3.7. Let P be a prime divisor of $\overline{I}$ for some ideal I in a Noetherian ring. Then there is a minimal prime q contained in P such that P/q is a prime divisor of $(\overline{I+q/q})$.

Proof: We may assume that R is local at P. Let $P = (\overline{I} : c)$. Since $c \notin \overline{I}$, I does not reduce (I,c). Thus Lemma 3.6 shows that for some minimal prime q, $I+q/q$ does not reduce $(I,c)+q/q$. Therefore $c+q \notin (\overline{I+q/q})$. We easily see that $(P/q)(c+q) \subseteq \overline{I} + q/q \subseteq (\overline{I+q/q})$, and the result follows.

LEMMA 3.8. Let $\mathcal{R}$ be the Rees ring of R with respect to an ideal I. If Q is a prime divisor of $\overline{t^{-n}\mathcal{R}}$ for some $n > 0$, then either $It \not\subseteq Q$ or $Q \cap R$ is a minimal prime in R.

Proof: By Lemma 3.7, there is a minimal prime q of $\mathcal{R}$ with Q/q a prime divisor of $(\overline{t^{-n}\mathcal{R}+q/q})$. We claim that $q \cap R[t^{-1}]$ is minimal in $R[t^{-1}]$. For this, let $S = \{1, t^{-1}, t^{-2}, \ldots\}$. Since t^{-1} is regular in $\mathcal{R}$, $q \cap S = \emptyset$. Now $R[t^{-1}]_S = \mathcal{R}_S = R[t^{-1}, t]$. Since q_S is minimal, the claim is obvious. Furthermore, if $p = q \cap R$, we see that p is minimal in R, that $q_S = pR[t^{-1}, t]$, and that the degree n component of q is $(I^n \cap p)t^n$ (with $I^n = R$ if $n \leq 0$). Since

$I^n/I^n \cap p \approx I^n + p/p$, we see that $\mathcal{R}/q \approx (R/p)[t^{-1}, (I+p/p)t] = \mathcal{R}'$, the Rees ring of R/p with respect to $I+p/p$. This isomorphism takes Q/q to a prime divisor Q' of $\overline{t^{-n}\mathcal{R}'}$. By Lemma 3.3, either $(I+p/p)t \not\subseteq Q'$ or $Q' \cap (R/p) = 0$. If $(I+p/p)t \not\subseteq Q'$ then clearly $It \not\subseteq Q$. If $Q' \cap R/p = 0$, then $Q/q \cap R/p = 0$, so that $Q \cap R = p$ is a minimal prime in R.

PROPOSITION 3.9. Let I be an ideal in a Noetherian ring R. Then $\bar{A}(I,1) \subseteq \bar{A}(I,2) \subseteq \ldots$ and this sequence eventually stabilizes.

Proof: If P is a minimal prime of R and $I \subseteq P$, then clearly P is in every term of our sequence. If $P \in \bar{A}(I,n)$ and P is not minimal in R, then the proof that $P \in \bar{A}(I,n+1)$ is identical to the first paragraph of the proof of Proposition 3.4 (using Lemma 3.8 instead of Lemma 3.3). Thus $\bar{A}(I,1) \subseteq \bar{A}(1,2) \subseteq \ldots$.

Now suppose $P \in \bar{A}(I,n)$. By Lemma 3.7, there is a minimal prime q of R with $q \subseteq P$ and P/q is in $\bar{A}(I+q/q,n)$. Thus $P/q \in \bar{A}^*(I+q/q)$ by Proposition 3.4. As R has only finitely many minimal primes q, and as $\bar{A}^*(I+q/q)$ is finite, we see that $\cup \bar{A}(I,n)$ is finite. Thus $\bar{A}(I,1) \subseteq \bar{A}(I,2) \subseteq \ldots$ eventually stabilizes.

DEFINITION. $\bar{A}^*(I)$ will denote the limit set of the above sequence.

Note: We will soon show that $P \in \bar{A}^*(I)$ if and only if for some minimal prime q, $P/q \in \bar{A}^*(I+q/q)$.

Remark: We point out that $\bar{A}^*(I)$ is well behaved with respect to localization. That is, $P \in \bar{A}^*(I)$ if and only if $P_S \in \bar{A}^*(I_S)$, S multiplicatively closed, $S \cap P = \emptyset$.

We have yet to show that $\bar{A}^*(I) \subseteq A^*(I)$. We choose to follow a path laid out by Dan Katz [Kz1] which touches many useful ideas.

PROPOSITION 3.10. Let $\mathcal{R}$ be the Rees ring of R with respect to I. The following statements are equivalent.

(i) $P \in \bar{A}^*(I)$

(ii) There is a prime $p \in \bar{A}^*(t^{-1}\mathcal{R})$ with $p \cap R = P$.

Proof: Since $\overline{I^n} = \overline{t^{-n}\mathcal{R}} \cap R$, any prime in $\overrightarrow{A}^*(I)$ lifts to a prime in $\overrightarrow{A}^*(t^{-1}\mathcal{R})$. Conversely, suppose p satisfies (ii), and write $p = (\overline{t^{-n}\mathcal{R}} : ct^m)$ with $c \in I^m$. Since the degree m component of $\overline{t^{-n}\mathcal{R}}$ is $(\overline{I^{n+m}} \cap I^m)t^m$, and since $p \cap R = P$, we have $P = (\overline{I^{n+m}} \cap I^m : c) = (\overline{I^{n+m}} : c)$ since $c \in I^m$. Clearly $c \notin \overline{I^{n+m}}$ since $P \neq R$. Thus $n+m > 0$, and we have $P \in \bar{A}(I, n+m) \subseteq \overrightarrow{A}^*(I)$.

LEMMA 3.11. Let I be an ideal in a Noetherian ring R. Then $\cap \overline{I^n}$, $n = 1,2,3,\ldots$ is the intersection of those minimal primes q such that $I + q \neq R$.

Proof: First assume that R is a domain. Let V be any D.V.R. containing R. Then $\overline{I^n} \subseteq \overline{I^n V} \subseteq \overline{I^n V} = I^n V$ since $I^n V$ is principal and V is integrally closed. Since $\cap I^n V = 0$, $\cap \overline{I^n} = 0$ as desired.

In general, say $q_1, \ldots, q_r, q_{r+1}, \ldots, q_s$ are the minimal primes of R with $I + q_i \neq R$ exactly when $i = 1,2,\ldots,r$. For these i we have $[(\cap \overline{I^n}) + q_i]/q_i \subseteq \cap \overline{[(I+q_i)/q_i]^n}$ (noting that $I + q_i$ is proper) which by the domain case shows that $\cap \overline{I^n} \subseteq q_1 \cap \ldots \cap q_r$. Now let $x \in q_1 \cap \ldots \cap q_r$. For any $n > 0$ and for any $j = r+1, \ldots, s$, we have $I^n + q_j = R$. Thus there is a $y \in q_{r+1} \cap \ldots \cap q_s$ and an $z \in I^n$ with $z + y = 1$. Therefore $x = zx + yx$. Now yx is in the radical of R, hence in $\cap \overline{I^n}$. As $zx \in I^n \subseteq \overline{I^n}$, we have $x \in \overline{I^n}$ for all $n > 0$.

LEMMA 3.12. Let I, P and q be ideals in a Noetherian ring R with q a minimal prime and with P a prime minimal over $I + q$. Then there is an $n \geq 1$ such that for any $m \geq n$ and any ideal J with $I^m \subseteq J \subseteq \overline{I^m}$, we have $P \in \mathrm{Ass}(R/J)$.

Proof: Localizing at P, we have $P^k \subseteq I + q$ for some $k > 0$. As q is minimal, there is an $x \notin q$ with $xq^n = 0$ for large n. Since R is local, the previous lemma shows that we may choose n large enough to assure that $x \notin \overline{I^n}$, as well.

Let $m \geq n$ and let $I^m \subseteq J \subseteq \overline{I^m}$. Now $P^{2mk} \subseteq (I+q)^{2m} \subseteq I^m + q^m$, so that $P^{2mk} x \subseteq I^m x + q^m x = I^m x \subseteq I^m \subseteq J$. Since $x \notin \overline{I^m}$, $x \notin J$. Thus P^{2mk} consists of zero divisors modulo J. As P is maximal, $P \in \mathrm{Ass}(R/J)$.

COROLLARY 3.13. Let I, P and q be as in the previous lemma. Then $P \in \overrightarrow{A}^*(I)$ and $P \in A^*(I)$.

Proof: Obvious.

Comparing this corollary to Proposition 1.14 the question arises whether we can weaken the above hypothesis to just q is a prime divisor of zero. The answer is no. In [FR] there is an example of a complete 2-dimensional local (R,M) such that every minimal prime has depth 2 and there also exists a prime divisor of zero, q, with depth $q = 1$. Choose $a \in M - q$. Then M is minimal over $aR + q$. However $M \notin \overline{A}^*(aR)$ since if it were, then by Lemma 3.7 (and Proposition 3.4) there would be a minimal prime p with $M/p \subseteq \overline{A}^*(aR+p/p)$. Since R is complete, R/p is quasi-unmixed and satisfies the altitude formula (see the Appendix). Our next lemma thus shows that height $M/p = 1$, contradicting that depth $p = 2$.

LEMMA 3.14. Let $a \neq 0$ in a Noetherian domain R which satisfies the altitude formula. Then $\overline{A}^*(aR) = \{P \text{ prime} \mid a \in P \text{ and height } P = 1\}$.

Proof: One inclusion is obvious. Suppose that $P \in \overline{A}^*(aR)$. By Lemma 3.2, $\overline{R}$ contains a height 1 prime p lying over P. Since R satisfies the altitude formula, height $P =$ height $p = 1$.

Lemma 3.14 will be greatly strengthened in Chapter 4.

LEMMA 3.15. Let T be a faithfully flat ring extension of R. If I is an ideal of R then $\overline{IT} \cap R = \overline{I}$.

Proof. Let $\mathcal{R} = R[t^{-1}, It]$ and $\mathcal{R}' = T[t^{-1}, It]$. We have $\overline{I} = \overline{t^{-1}\mathcal{R}} \cap R$ and $\overline{IT} = \overline{t^{-1}\mathcal{R}'} \cap T$, so that it will suffice to show that $\overline{t^{-1}\mathcal{R}'} \cap \mathcal{R} = \overline{t^{-1}\mathcal{R}}$. Therefore, since $\mathcal{R}'$ is a faithfully flat extension of $\mathcal{R}$, it will be enough to prove the statement of the lemma in the special case that $I = bR$ is principal. For this, suppose that $x \in \overline{bT} \cap R$. We easily see that for some n, $x^n \in (bx^{n-1}T + b^2x^{n-2}T + \ldots + b^nT) \cap R = (bx^{n-1}, \ldots, b^n)T \cap R = (bx^{n-1}, \ldots, b^n)R$, since T is a faithfully flat extension of R. This clearly shows that $x \in \overline{bR}$.

PROPOSITION 3.16. Let (R,M) be a local ring with completion R^*. Let I be an ideal of R. Then $P \in \overrightarrow{A}^*(I)$ iff and only if there is a $P^* \in \overrightarrow{A}^*(IR^*)$ with $P^* \cap R = P$.

Proof: By Lemma 3.15, we see that any prime in $\overrightarrow{A}^*(I)$ lifts to a prime in $\overrightarrow{A}^*(IR^*)$. Thus suppose that $P^* \in \overrightarrow{A}^*(IR^*)$ and let $P = P^* \cap R$. Also let $\mathcal{R} = R[t^{-1}, It]$ and $\mathcal{R}^* = R^*[t^{-1}, IR^*t]$. By Proposition 3.10, there is a $p^* \in \overrightarrow{A}^*(t^{-1}\mathcal{R}^*)$ with $p^* \cap R^* = P^*$. By Lemma 3.7, for some minimal prime $q^* \subseteq p^*$ we have $p^*/q^* \in \overrightarrow{A}^*(t^{-1}\mathcal{R}^* + q^*/q^*)$. Now R^* being complete implies that $R^*/q^* \cap R^*$ satisfies the altitude formula, as does its finitely generate extension $\mathcal{R}^*/q^*$. By Lemma 3.14, we see that p^*/q^* has height 1. Thus p^* is minimal over $t^{-1}\mathcal{R}^* + q^*$. Since for any $m > 0$, $\overline{t^{-m}\mathcal{R}^*} \subseteq \overline{(t^{-m}\mathcal{R})}\mathcal{R}^* \subseteq \overline{t^{-m}\mathcal{R}^*}$, we see by Lemma 3.12 that for large m, p^* is a prime divisor of $\overline{(t^{-m}\mathcal{R})}\mathcal{R}^*$. Now $R \subseteq R^*$ is a flat extension. Thus so is $\mathcal{R} \subseteq \mathcal{R}^*$. Therefore $p^* \cap \mathcal{R}$ is a prime divisor of $\overline{t^{-m}\mathcal{R}}$. By Proposition 3.10, $P = (p^* \cap \mathcal{R}) \cap R$ is in $\overrightarrow{A}^*(I)$.

PROPOSITION 3.17. Let I be an ideal in a Noetherian ring R. Then $\overrightarrow{A}^*(I) \subseteq A^*(I)$. In fact, if $P \in \overrightarrow{A}^*(I)$ then either P is minimal in R or $P \in B^*(I)$.

Proof: Assume that $P \in \overrightarrow{A}^*(I)$ is not minimal, and localize at P. By Proposition 3.16, we see that $P^* \in \overrightarrow{A}^*(IR^*)$. Take p^* and q^* as in the proof of that proposition. Since p^* is minimal over $t^{-1}\mathcal{R}^* + q^*$, by Corollary 3.13, $p^* \in A^*(t^{-1}\mathcal{R}^*)$. As $\mathcal{R} \subseteq \mathcal{R}^*$ is a flat extension, $p^* \cap \mathcal{R} \in A^*(t^{-1}\mathcal{R})$. Now t^{-1} is regular in $\mathcal{R}$, and so we easily see that $p^* \cap \mathcal{R}$ is a prime divisor of $t^{-1}\mathcal{R}$. Also, in the proof of Proposition 3.16 we saw that $p^* \cap \mathcal{R}$ was a prime divisor of $\overline{t^{-m}\mathcal{R}}$ for large m. Since $P = (p^* \cap \mathcal{R}) \cap R$ is not minimal, by Lemma 3.8, $It \not\subseteq p^* \cap \mathcal{R}$. Proposition 1.15 tells us that $P \in B^*(I)$. Finally, since any minimal prime of R containing I is in $A^*(I)$, we see that $\overrightarrow{A}^*(I) \subseteq A^*(I)$.

PROPOSITION 3.18. Let I be an ideal in a Noetherian ring, and let $\mathcal{R} = R[t^{-1}, It]$. The following statements are equivalent.

i) $P \in \overrightarrow{A}^*(I)$.

ii) There is a $p \in \overrightarrow{A}^*(t^{-1}\mathcal{R})$ with $p \cap R = P$.

iii) There is a prime divisor p' of $t^{-1}\overline{\mathcal{R}}$ with $p' \cap R = P$. ($\overline{\mathcal{R}}$ is the integral closure of $\mathcal{R}$ in its total quotient ring.)

iv) There is a minimal prime q of R with $q \subseteq P$ and $P/q \in \overline{A}^*(I+q/q)$.

Also, if R is local with completion R^*, the above are equivalent to

v) There is a $P^* \in \overline{A}^*(IR^*)$ with $P^* \cap R = P$.

Proof: We already have (i) $\Longleftrightarrow$ (ii) $\Longleftrightarrow$ (v). By Lemma 3.7 and Proposition 3.4, we have (i) $\Rightarrow$ (iv). We now show (iv) $\Rightarrow$ (i), proceeding in three steps. First, assume that I is principal and that R/q satisfies the altitude formula. Then $I+q/q$ is principal, and by Lemma 3.14 we have that height $P/q = 1$ (or else $I+q/q = 0$ in which case $P = q$ is minimal, making (i) trivially true). Thus P is minimal over $I+q$, and so $P \in \overline{A}^*(I)$ by Corollary 3.13. For step two, we only assume that I is principal. Now we may localize at P, and assume that R is local. By (i) $\Rightarrow$ (v) and the fact that $(R/q)^* \approx R^*/qR^*$, we have that $P^*/qR^* \in \overline{A}^*(IR^* + qR^*/qR^*)$. Now by (i) $\Rightarrow$ (iv) and the fact that minimal primes of R^*/qR^* have the form q^*/qR^* with q^* a minimal prime of R^*, we see that there is such a q^* with $P^*/q^* \in \overline{A}^*(IR^* + q^*/q^*)$. Since I is principal and R^*/q^* satisfies the altitude formula, by the first step we have that $P^* \in \overline{A}^*(IR^*)$. By (v) $\Rightarrow$ (i) we have $P \in \overline{A}^*(I)$.

For the final step, we consider the general case. Let $\mathcal{R}_0$ be the Rees ring of R/q with respect to $I+q/q$. Let $q^+ = qR[t^{-1},t] \cap \mathcal{R}$, which is a minimal prime of $\mathcal{R}$, and notice that $\mathcal{R}/q^+ \approx \mathcal{R}_0$. Since $P/q \in \overline{A}^*(I+q/q)$, (i) $\Rightarrow$ (ii) shows that there is a prime p of $\mathcal{R}$ with $p/q^+ \in \overline{A}^*(t^{-1}\mathcal{R}+q^+/q^+)$, and with $p/q^+ \cap R/q = P/q$, so that $p \cap R = P$. Step two of our argument shows that $p \in \overline{A}^*(t^{-1}\mathcal{R})$, and so by (ii) $\Rightarrow$ (i) we have $P \in \overline{A}^*(I)$. This proves that (iv) $\Rightarrow$ (i).

For (ii) $\Rightarrow$ (iii) we observe that $\overline{t^{-n}\mathcal{R}} = t^{-n}\overline{\mathcal{R}} \cap R$ so that any $p \in \overline{A}^*(t^{-1}\mathcal{R})$ can be lifted to a prime divisor p' of $t^{-n}\overline{\mathcal{R}}$ for some n. As t^{-1} is regular in $\overline{\mathcal{R}}$, p' is also a prime divisor of $t^{-1}\overline{\mathcal{R}}$. Finally, for (iii) $\Rightarrow$ (ii), let p' be a prime divisor of $t^{-1}\overline{\mathcal{R}}$. Then height $p' = 1$. Let q' be a minimal prime of $\overline{\mathcal{R}}$ contained in p'. Then $q = q' \cap \mathcal{R}$ is a minimal prime of $\mathcal{R}$ contained in $p = p' \cap \mathcal{R}$.

Since height $p'/q' = 1$, Proposition 1.7 shows that $p/q \in \overrightarrow{A}^*(t^{-1}\mathcal{R} + q/q)$. By (iv) $\Rightarrow$ (i), $p \in \overrightarrow{A}^*(t^{-1}\mathcal{R})$, so that (iii) $\Rightarrow$ (ii).

By now, our next result, due to Ratliff, is straightforward. However, it motivated many of the ideas in the subject (such as Proposition 1.7) and so warrants mention.

PROPOSITION 3.19. [R6] Let (R,M) be a local domain with completion R^* and integral closure $\overline{R}$. The following are equivalent

(i) R^* contains a depth 1 minimal prime.

(ii) $M \in \overrightarrow{A}^*(I)$ for every ideal $I \neq 0$.

(iii) $M \in \mathrm{Ass}(R/\overline{(a)})$ for some $a \neq 0$.

(iv) $\overline{R}$ contains a height 1 maximal prime.

(v) There is an $n > 0$ such that for every ideal $0 \neq I \subseteq \overline{M^n}$, $M \in \mathrm{Ass}(R/I)$.

(vi) There is an $n > 0$ such that for every ideal $0 \neq I \subseteq M^n$, $M \in \mathrm{Ass}(R/\overline{I})$.

Proof: (i) $\Rightarrow$ (ii). Let $I \neq 0$ and let M^* be the maximal prime of R^*. Also let q^* be a depth 1 minimal prime of R^*. Then M^*/q^* is minimal over $IR^* + q^*/q^*$, so that $M^*/q^* \in \overrightarrow{A}^*(IR^* + q^*/q^*)$. By Proposition 3.18 (iv) $\Rightarrow$ (i) and (v) $\Rightarrow$ (i), we have $M \in \overrightarrow{A}^*(I)$.

(ii) $\Rightarrow$ (iii) is obvious.

(iii) $\Rightarrow$ (i) by Proposition 3.4, we have $M \in \overrightarrow{A}^*(aR)$. By Proposition 3.18 used twice, there is a minimal prime q^* of R^* with $M^*/q^* \in \overrightarrow{A}^*(aR^* + q^*/q^*)$. Now R^*/q^* satisfies the altitude formula, so Lemma 3.14 shows that depth $q^* = 1$.

(iii) $\Rightarrow$ (iv) is by Lemma 3.2.

(iv) $\Rightarrow$ (ii) is by Proposition 3.5.

(iv) $\Rightarrow$ (v) is by Proposition 1.7.

(v) $\Rightarrow$ (vi) is immediate since $I \subseteq M^n$ implies $\overline{I} \subseteq \overline{M^n}$.

(vi) $\Rightarrow$ (iii) is obvious.

The following proposition and its applications are due to K. Whittington [W].

PROPOSITION 3.20. Let $I = (a_1, \ldots, a_n)$ be a nonzero ideal in a Noetherian domain R. Let $I \subseteq P$ with P prime. The following are equivalent.

(i) $P \in \overrightarrow{A}^*(I)$.

(ii) For some a_i, there is a prime Q of $A = R[a_1/a_i, \ldots, a_n/a_i]$ with $Q \in \overrightarrow{A}^*(a_iA)$ and $Q \cap R = P$.

(iii) For some a_i, the integral closure of $R[a_1/a_i, \ldots, a_n/a_i]$ contains a height 1 prime lying over P.

Proof: (ii) $\Leftrightarrow$ (iii) easily follows from Proposition 3.19 after the appropriate localization.

(i) $\Rightarrow$ (ii). Let $\mathcal{R} = R[t^{-1}, It]$. By Proposition 3.10, there is a $p \in \overrightarrow{A}^*(t^{-1}\mathcal{R})$ with $p \cap R = P$. By Proposition 3.19, there is a height 1 prime p' of $\overline{\mathcal{R}}$ with $p' \cap \mathcal{R} = p$. Let $(V,N) = (\overline{\mathcal{R}}_{p'}, p'_{p'})$, which is a D.V.R. By Lemma 3.3, $It \not\subseteq p$. Thus pick a_i such that $a_it \notin p$. With $A = R[a_1/a_i, \ldots, a_n/a_i]$, we have $A \subseteq V$, since $a_j/a_i = a_jt/a_it$ and $a_it \in V - N$. Let $\mathcal{A} = A[t^{-1}, a_it]$ be the Rees ring of A with respect to a_iA. Clearly $A \subseteq \mathcal{A} \subseteq \overline{\mathcal{A}} \subseteq V$. Also $\mathcal{R} \subseteq \mathcal{A}$, since $a_jt = (a_j/a_i)a_it$. Thus $\overline{\mathcal{R}} \subseteq \overline{\mathcal{A}} \subseteq V$. Since $N = p'_{p'}$, obviously $N \cap \overline{\mathcal{A}}$ is a height 1 prime of $\overline{\mathcal{A}}$ containing t^{-1}. Proposition 3.19 shows that $N \cap \mathcal{A} \in \overrightarrow{A}^*(t^{-1}\mathcal{A})$. By Proposition 3.10, $N \cap A \in \overrightarrow{A}^*(a_iA)$. Letting $Q = N \cap A$, we see that (i) $\Rightarrow$ (ii).

(ii) $\Rightarrow$ (i). Let a_i, A and Q be as in (ii). By Proposition 3.10, $\mathcal{A} = A[t^{-1}, a_it]$ contains a prime $q \in \overrightarrow{A}^*(t^{-1}\mathcal{A})$ with $q \cap A = Q$. By Lemma 3.3, $a_it \notin q$. Now Proposition 3.19 shows that $\overline{\mathcal{A}}$ contains a height 1 prime q' lying over q. Let (W,K) be the D.V.R. $(\overline{\mathcal{A}}_{q'}, q'_{q'})$. Now $\mathcal{R} = [t^{-1}, It] \subseteq \mathcal{A}$, so that W is a localization of an integral extension of a finitely generated extension of $\mathcal{R}$. Clearly the transcendence degree of W/K over $\mathcal{A}/q$ is 0. Now $\mathcal{A}$ is obtained by adjoining $a_j/a_i = a_jt/a_it$ to $\mathcal{R}$, and $a_it \notin q$. Thus the transcendence degree of $\mathcal{A}/q$ over $\mathcal{R}/q \cap \mathcal{R}$ is 0. Therefore, the transcendence degree of W/K over $\mathcal{R}/q \cap \mathcal{R}$

is 0. By Lemma 3.1, height $K \cap \overline{\mathcal{R}} = 1$. Since $t^{-1} \in K \cap \overline{\mathcal{R}}$, Proposition 3.19 shows that $q \cap \mathcal{R} = K \cap \mathcal{R} \in \overrightarrow{A}^*(t^{-1}\mathcal{R})$. Finally, $P = (q \cap \mathcal{R}) \cap R$, and Proposition 3.10 shows that $P \in \overrightarrow{A}^*(I)$.

PROPOSITION 3.21. Let $I \neq 0$ be an ideal in a Noetherian ring R, and let X be an indeterminate. Then $\overrightarrow{A}^*(IR[X]) = \{PR[X] \mid P \in \overrightarrow{A}^*(I)\}$.

Proof: First assume that R is a domain, and let P be a prime containing I. Let $I = (a_1, \ldots, a_n, b)$, and let $A = R[a_1/b, \ldots, a_n/b]$. Clearly $\overline{A}$ contains a height 1 prime lying over P if and only if $\overline{A}[X]$ contains a height 1 prime lying over $PR[X]$. Thus Proposition 3.20 easily shows that $P \in \overrightarrow{A}^*(I)$ if and only if $PR[X] \in \overrightarrow{A}^*(IR[X])$.

Now let Q be prime in $R[X]$, with $Q \cap R = P$, but $Q \neq PR[X]$. To complete the domain case, we must show that $Q \notin \overrightarrow{A}^*(IR[X])$. Suppose, contrarily, that $Q \in \overrightarrow{A}^*(IR[X])$. We may assume that R is local at P, so that $Q = (P, f(X))$ with $f(X)$ a monic polynomial. By Proposition 3.20, we may write $I = (a_1, \ldots, a_n, b)$ such that if $A = R[a_1/b, \ldots, a_n/b]$ then $\overline{A}[X] = \overline{R[X, a_1/b, \ldots, a_n/b]}$ contains a height 1 prime q lying over Q. Let $p = q \cap \overline{A}$. Since $q \cap R = Q \cap R = P$, $p \cap R = P$ so that $p \neq 0$. Now since height $q = 1$, we must have $q = p\overline{A}[X]$. However $f(X) \in Q \in q$ and $f(X)$ is monic so that $f(X) \notin p\overline{A}[X]$. This contradiction completes the domain case.

For the general case, we have $Q \in \overrightarrow{A}^*(IR[X])$ if and only if $Q/q \in \overrightarrow{A}^*(IR[X]+q/q)$ for some minimal prime q of $R[X]$, by Proposition 3.18. Now q has form $q = pR[X]$ with p a minimal prime in R. Since the isomorphism $R[x]/q \approx R/p[X]$ carries $IR[X]+q/q$ to $(I+p/p)R/p[X]$, and since by the domain case just discussed, the primes in $\overrightarrow{A}^*((I+p/p)R/p[X])$ have the form $(P/p)R/p[X]$ for $P/p \in \overrightarrow{A}^*(I+p/p)$, we see that $Q \in \overrightarrow{A}^*(IR[X])$ if and only if Q/q is isomorphic to $(P/p)R/p[X]$ for some $P/p \in \overrightarrow{A}^*(I+p/p)$. Finally $P/p \in \overrightarrow{A}^*(I+p/p)$ if and only if $P \in \overrightarrow{A}^*(I)$ by Proposition 3.18. Because Q/q is isomorphic to $(P/p)R/p[X]$ implies $Q = PR[X]$, we are done.

PROPOSITION 3.22. Let $R \subseteq T$ be an integral extension of Noetherian rings. Let I be an ideal in R. If $P \in \overrightarrow{A}^*(I)$ then there is a $Q \in \overrightarrow{A}^*(IT)$ with $Q \cap R = P$. If also every minimal prime of T contracts to a minimal prime of R, then $Q \in \overrightarrow{A}^*(IT)$ implies $Q \cap R \in \overrightarrow{A}^*(I)$.

Proof: First suppose that R and T are domains, and let P be prime in R. We must show that $P \in \overrightarrow{A}^*(I)$ if and only if there is a $Q \in \overrightarrow{A}^*(IT)$ with $Q \cap R = P$. If $P \in \overrightarrow{A}^*(I)$, by Proposition 3.20 we may write $I = (a_1, \ldots, a_n, b)$ such that $\overline{R[Ib^{-1}]}$ contains a height 1 prime p lying over P. Now $\overline{R[Ib^{-1}]} \subseteq \overline{T[ITb^{-1}]}$ is an integral extension. Thus there is a height 1 prime q of $\overline{T[ITb^{-1}]}$ lying over p. Let $Q = q \cap T$. By Proposition 3.20, $Q \in \overrightarrow{A}^*(IT)$. Also $Q \cap R = P$. The converse of the domain case is similar, except it needs the additional fact that $\overline{R[Ib^{-1}]} \subseteq \overline{T[ITb^{-1}]}$ satisfies going down, so that height 1 primes contract to height 1 primes.

In the general case, suppose that $P \in \overrightarrow{A}^*(I)$. By Proposition 3.18 (i) $\Rightarrow$ (iv), $P/p \in \overrightarrow{A}^*(I+p/p)$ for some minimal prime p of R. Lift p to a prime q of T, which must be minimal. Applying the domain case to $R/p \subseteq T/q$, we find a $Q/q \in \overrightarrow{A}^*(IT+q/q)$ with $Q/q \cap R/p = P/p$. Thus $Q \cap R = P$ and by Proposition 3.18, $Q \in \overrightarrow{A}^*(IT)$. If every minimal prime of T contracts to a minimal prime of R, the reverse works similarly.

Remarks: (a) The condition on minimal primes above is required. It is easy to find $R \subseteq T$ as above with R a domain but q a minimal prime in T with $q \cap R \neq 0$. Now $q \in \overrightarrow{A}^*(0T)$ but $q \cap R \notin \overrightarrow{A}^*(0R)$.

(b) The above result fails for $A^*(I)$. To see this, let (R,M) be a 2-dimensional local domain which satisfies the altitude formula, but which is not Cohen-Macaulay, and let $\overline{R}$ be Noetherian. (Such a situation is easily constructed.) For $0 \neq a \in M$, we have $M \in A^*(aR)$ since R is not Cohen-Macaulay. Suppose N is a prime of $\overline{R}$ lying over M. In view of the altitude formula, we have height N = height $M = 2$. As $\overline{R}$ is a Krull domain, N cannot be a prime divisor of $a^n\overline{R}$ for any n. Thus $N \notin A^*(a\overline{R})$.

Let R be a Noetherian domain and $I \neq 0$ an ideal. If $P \in \overline{A}^*(I)$, then Proposition 3.20 tells us that for some $0 \neq b \in I$, $\overline{R[Ib^{-1}]}$ contains a height 1 prime Q lying over P. Let (V,N) be the D.V.R. $(\overline{R[Ib^{-1}]}_Q, Q_Q)$.

LEMMA 3.23. Let R, I, P, b, Q and V be as above. If $J \neq 0$ is an ideal of R with $JV \subseteq IV$ and $(J : I)V = (JV : IV)$, then $P \in \overline{A}^*(J)$.

Proof: Since $JV \subseteq IV$, $J \subseteq IV \cap R \subseteq N \cap R = P$. Also $JV \subseteq IV$ are both principal ideals of V and so $JV = (JV : IV)IV$. Note that $IV = bV$. If we choose $c \in J$ and $d \in (J : I)$ such that $JV = cV$ and $(J : I)V = dV$, then $cV = JV = (JV : IV)IV = (J : I)VIV = dbV$. Since $J \subseteq JV = cV = dbV$, $J(db)^{-1} \subseteq V$. Also, since $b \in I$ and $d \in (J : I)$, $bd \in J$. Furthermore, $Id \subseteq I(J : I)$ so that $Ib^{-1} \subseteq I(J : I)(bd)^{-1} \subseteq J(bd)^{-1}$. Thus $R[Ib^{-1}] \subseteq R[J(bd)^{-1}] \subseteq V$, so that $\overline{R[Ib^{-1}]} \subseteq \overline{R[J(bd)^{-1}]} \subseteq V$. Since $N \cap \overline{R[Ib^{-1}]} = Q$ has height 1, clearly $N \cap \overline{R[J(bd)^{-1}]}$ has height 1. Since this prime lies over P, and because we already have $J \subseteq P$ and $bd \in J$, Proposition 3.20 shows that $P \in \overline{A}^*(J)$.

If $P \in \overline{A}^*(I)$ and if $I \subseteq J \subseteq \overline{I}$, then obviously $P \in \overline{A}^*(J)$. The next corollary goes a little further, since $\overline{I} \subseteq IV \cap R$.

COROLLARY 3.24. Let R, I, P and V be as in Lemma 3.23. If J is an ideal of R with $I \subseteq J \subseteq IV \cap R$, then $P \in \overline{A}^*(J)$.

Proof: Obviously $IV = JV$ and $(J : I)V = RV = V = (JV : IV)$.

COROLLARY 3.25. Let $P \in \overline{A}^*(I)$. Then there is an $n > 0$ such that if J is any ideal with $I \subseteq J \subseteq I + \overline{P^n}$, then $P \in \overline{A}^*(J)$.

Proof: First assume that R is a domain. By Proposition 3.20, we can find b, Q and V as in Lemma 3.23. Since V is a D.V.R., for some n we have $P^n V \subseteq IV$. Now $\overline{P^n} \subseteq P^n V \subseteq IV$. Thus for J as above, $I \subseteq J \subseteq IV \cap R$, and so $P \in \overline{A}^*(J)$ by Corollary 3.24. The general case follows by Proposition 3.18 ($(i) \Leftrightarrow (iv)$) and the fact that $\overline{P^n} + q/q \subseteq \overline{(P/q)^n}$ for $q \subseteq P$ with q a minimal prime.

PROPOSITION 3.26. Let $I \subseteq P$ be ideals of R with P prime. Assume $\dim R > 0$. The following are equivalent.

(i) $P \in \overrightarrow{A}^*(I)$.

(ii) $P \in \overrightarrow{A}^*(IJ)$ for any ideal J with height $J > 0$.

(iii) $P \in \overrightarrow{A}^*(Ic)$ for any element c not contained in any minimal prime.

(iv) There exists an element c not in any minimal prime, with $P \in \overrightarrow{A}^*(Ic)$.

Proof: (ii) $\Rightarrow$ (iii) is immediate, as is (iii) $\Rightarrow$ (iv) since $\dim R > 0$. We will prove (i) $\Rightarrow$ (ii) and (iv) $\Rightarrow$ (i) for R a domain. The general case is then straightforward using Proposition 3.18 ((i) $\Leftrightarrow$ (iv)).

Assume R is a domain, and let $P \in \overrightarrow{A}^*(I)$. Let J be any nonzero ideal. Since $P \in \overrightarrow{A}^*(I)$, Proposition 3.20 shows the existence of a b, Q and V as in Lemma 3.23. As IV is principal, $(IJV : IV) = (JV : V) = JV$, so that $(IJ : I)V \subseteq (IJV : IV) = JV \subseteq (IJ : I)V$. Thus $(IJ : I)V = (IJV : IV)$. As $IJV \subseteq IV$, Lemma 3.23 shows that $P \in \overrightarrow{A}^*(IJ)$. Thus (i) $\Rightarrow$ (ii). For (iv) $\Rightarrow$ (i), suppose $c \neq 0$ and $P \in \overrightarrow{A}^*(Ic)$. Let $I = (a_1, \ldots, a_n)$. By Proposition 3.20, for some i, $\overline{R[a_1c/a_ic, \ldots, a_nc/a_ic]}$ contains a height 1 prime lying over P. As this ring equals $\overline{R[a_1/a_i, \ldots, a_n/a_i]}$, that proposition shows $P \in \overrightarrow{A}^*(I)$.

In Chapter 4, we will give an example where $P \in \overrightarrow{A}^*(IJ)$, but $P \notin \overrightarrow{A}^*(I)$ and $P \notin \overrightarrow{A}^*(J)$.

CHAPTER IV: A Characterization of $\overrightarrow{A}^*(I)$

In this chapter we investigate a fundamental characterization of $\overrightarrow{A}^*(I)$ for ideals in a Noetherian ring which is locally quasi-unmixed (see Appendix). Recall that if I is an ideal in a local ring (R,M), then $\ell(I)$, the analytic spread of I, can be characterized in various ways. If $f(n)$ is the minimal number of generators of I^n, then there is a polynomial $P(X)$ such that $P(n) = f(n)$ for all large n (the Hilbert polynomial), and $\ell(I) = \deg P + 1$. If $\mathcal{R} = R[t^{-1}, It]$ and if $\mathcal{M} = \ldots + Rt^{-2} + Rt^{-1} + M + It + I^2t^2 + \ldots$, then $\ell(I)$ equals the height of $\mathcal{M}/(t^{-1},M)\mathcal{R}$. Finally, if R/M is infinite, so that I has a minimal reduction, then $\ell(I)$ is the size of a minimal basis of a minimal reduction of I. Recall also that for a Noetherian domain R, being locally quasi-unmixed is equivalent to satisfying the altitude formula.

PROPOSITION 4.1. [M4] Let I be an ideal in a Noetherian ring R and let P be a prime containing I. If height $P = \ell(I_P)$, then $P \in \overrightarrow{A}^*(I)$. If R_P is quasi-unmixed, then the converse is also true.

Proof: We first treat the case that R is a domain. We localize at P. Since $\ell(I) = \ell(IR(X))$ (note: $R(X) = R[X]_{PR[X]}$), by Proposition 3.21 we may assume that R/P is infinite. Let J be a minimal reduction of I. Since $\overline{I^m} = \overline{J^m}$ for all m, obviously $\overrightarrow{A}^*(I) = \overrightarrow{A}^*(J)$, and we may take $I = J = (a_1, \ldots, a_n)$ with $\ell(I) = n$. Now $a_1, \ldots, a_n$ are analytically independent. Letting $A = R[a_1/a_n, \ldots, a_{n-1}/a_n]$ we have that PA is a prime in A with height $PA \leq$ height $P - n + 1$. Suppose now that $\ell(I) = n =$ height P. Then height $PA = 1$ so that $\overline{A}$ has a height 1 prime lying over P. Therefore, by Proposition 3.20 we now see that $P \in \overrightarrow{A}^*(I)$. Conversely suppose that $P \in \overrightarrow{A}^*(I)$ and that (R,P) is quasi-unmixed, so that it satisfies the altitude formula. By Proposition 3.20 we have that $\overline{A}$ contains a height 1 prime q lying over P. As A is finitely generated over R, A also satisfies the altitude formula, and so height$(q \cap A) =$ height $q = 1$. Since PA is clearly contained in $q \cap A$, we have $PA = q \cap A$. Thus height $PA = 1$. However, the altitude formula also shows that height $PA =$ height $P - n + 1$. Thus height $P = n = \ell(I)$.

This completes the domain case. Before doing the general case, we require two lemmas.

LEMMA 4.2. Let I be an ideal in a local ring (R,M). For any minimal prime q, $\ell(I+q/q) \leq \ell(I)$, and equality holds for some such q.

Proof: Since the minimal number of generators of I^n+q/q does not exceed the minimal number of generators of I^n, the characterization of analytic spread in terms of Hilbert polynomials gives the inequality. Now in $\mathcal{R}$, the Rees ring of R with respect to I, let $N = \ldots + Rt^{-1} + M + It + I^2t^2 + \ldots$ and $J = \ldots + Rt^{-1} + M + MIt + MI^2t^2 + \ldots$ so that $\ell(I) = \text{height}(N/J)$. If this number is n, let $J \subset P_0 \subset P_1 \subset \ldots \subset P_n = N$ with P_i prime. Let Q be a minimal prime of $\mathcal{R}$ with $Q \subseteq P_0$. If $q = Q \cap R$ it is straightforward to verify that q is a minimal prime of R and that $Q = \ldots + qt^{-1} + q + (I \cap q)t + (I^2 \cap q)t^2 + \ldots$. Clearly N/Q modulo $J+Q/Q$ has height n. Now if $\mathcal{R}'$ is the Rees ring of R/q with respect to $I+q/q$, and if $N' = \ldots + R/q\ t^{-1} + M/q + (I+q/q)t + (I^2+q/q)t^2 + \ldots$ and $J' = \ldots + R/q\ t^{-1} + M/q + (MI+q/q)t + (MI^2+q/q)t^2 + \ldots$ then $\ell(I+q/q) = \text{height}\ N'/J'$. However $\mathcal{R}/Q \approx \mathcal{R}'$, the isomorphism taking N/Q to N' and $J+Q/Q$ to J'. Thus height $N'/J' = n = \ell(I)$.

LEMMA 4.3. (Rees) Let I be an ideal in a local ring (R,M). Then $\ell(I) \leq \text{height}\, M$.

Proof: By the previous lemma, for some minimal prime q, $\ell(I) = \ell(I+q/q)$. Since height $M/q \leq$ height M, we may assume R is a domain. We may also assume that R/M is infinite, and let $(a_1, \ldots, a_n)$ be a minimal reduction of I, with $n = \ell(I)$. Thus $a_1, \ldots, a_n$ are analytically independent. Applying the altitude inequality to the primes M and $MR[a_1/a_n, \ldots, a_{n-1}/a_n]$, since the transcendence degree of $R]a_1/a_n, \ldots, a_{n-1}/a_n]$ modulo $MR[a_1/a_n, \ldots, a_{n-1}/a_n]$ over R/M is $n-1$, we see that height $M \geq$ height $MR[a_1/a_n, \ldots, a_{n-1}/a_n] + (n-1) \geq n = \ell(I)$.

Proof of Proposition 4.1: (continued) In general, suppose that height P equals $\ell(I_P)$. We may assume that R is local at P. By Lemma 4.2, pick a minimal prime q of R with $\ell(I) = \ell(I+q/q)$. By Lemma 4.3, we see that $\ell(I+q/q) = \text{height}\ P/q$. By the domain case, $P/q \in \overrightarrow{A}^*(I+q/q)$, so that by Proposition 3.18, $P \in \overrightarrow{A}^*(I)$.

Finally suppose that $P \in \overline{A}^*(I)$ and that (R,P) is quasi-unmixed. By Proposition 3.18, for some minimal prime, $P/q \in \overline{A}^*(I+q/q)$. By the domain case, height $P/q = \ell(I+q/q)$. As R is quasi-unmixed, height $P/q =$ height P. By Lemmas 4.2 and 4.3, height $P = \ell(I)$.

COROLLARY 4.4. Let (R,M) be a local ring with completion R^*. If I is an ideal of R, then $M \in \overline{A}^*(I)$ if and only if for some minimal prime q^* of R^*, $\ell(IR^*+q^*/q^*) =$ height M^*/q^*.

Proof: By (i) $\Rightarrow$ (v) and (i) $\Rightarrow$ (iv) of Proposition 3.18, $M \in \overline{A}^*(I)$ if and only if for some minimal q^* of R^*, $M^*/q^* \in \overline{A}^*(IR^*+q^*/q^*)$. As R^*/q^* is quasi-unmixed, we invoke Proposition 4.1 to complete the proof.

Let R be locally quasi-unmixed. Then Proposition 4.1 shows that $P \in \overline{A}^*(I)$ if and only if height $P = \ell(I_P)$. This property characterizes locally quasi-unmixed rings. That is, if for all I in R, $P \in \overline{A}^*(I)$ if and only if height $P = \ell(I_P)$, then R is locally quasi-unmixed. (In fact we can restrict ourselves to ideals in the principal class, i.e. for which the height of I equals the minimal number of generators of I.) We could prove this now, using Corollary 4.4. Instead, we defer it until Chapter 5, when we will have the machinery to give a very brief proof.

EXAMPLE. We give the previously promised example of $P \in \overline{A}^*(IJ)$ with $P \notin \overline{A}^*(I)$ and $P \notin \overline{A}^*(J)$. Let (R,P) be a 3-dimensional quasi-unmixed local domain. Let a,b,c be a system of parameters. Let $I = (a,c)$ and $J = (b,c)$. Since $\ell(I)$ does not exceed the minimal number of generators of I, $\ell(I) \leq 2 <$ height P. By Proposition 4.1, $P \notin \overline{A}^*(I)$. Similarly $P \notin \overline{A}^*(J)$. Now $IJ = (ab,ac,bc,c^2)$, and since $ab/c^2 = (a/c)(b/c)$, we have $R[IJ(c^2)^{-1}] = R[a/c,b/c]$. Since a,b,c is a system of parameters, $PR[a/c,b/c]$ is a height 1 prime ideal. Thus $\overline{R[IJ(c^2)^{-1}]}$ contains a height 1 prime lying over P, and so $P \in \overline{A}^*(IJ)$ by Proposition 3.20

EXAMPLE. Let R equal $K[X,Y,Z,W]/(XY-ZW)$ localized at $(X,Y,Z,W)/(XY-ZW)$. With primes denoting images in R, let $P = (X',Z')$. We will show that $\overline{A}^*(P) = \{P\}$. Let $p \in \overline{A}^*(P)$. Since R satisfies the altitude formula, height $p = \ell(P_p)$ by

Proposition 4.1. Since $\ell(P_p)$ does not exceed the minimal number of generators of P_p, height $p \leq 2$. Thus $p \neq (X',Y',Z',W')$. Since $P \subseteq p$ we must have either $Y' \notin p$ or $W' \notin p$. Since $X'Y' = Z'W'$, we therefore easily see that P_p is generated by either X' or Z'. Thus height $p = \ell(P_p) = 1$, so that $p = P$.

Remark: The above example is quite interesting. P is not principal since if it were, then (X',Y',Z',W') could be generated by three elements. Since $XY - ZW \in (X,Y,Z,W)^2$, this would give (X,Y,Z,W) generated by three elements. Thus P is not principal. As R is normal [S2, Theorem 1], $\ell(P) > 1$ (since if $\ell(P) = 1$ with R normal, there is an $a \in P$ with $aR \subseteq P \subseteq \overline{aR} = aR$). Thus $\ell(P) = 2$ since P has two generators. However, the argument used above shows that $\ell(P_p) = 1$ for any height 2 prime p containing P.

LEMMA 4.5. Let (R,M) be a local domain with integral closure $\overline{R}$, and let $I \neq 0$ be an ideal of R. Then $\ell(I) = 1$ if and only if $I\overline{R}$ is principal.

Proof: By going to $R(X)$ if necessary, we may assume that R/M is infinite. Suppose that $\ell(I) = 1$. Then for some $a \in I$, $\overline{I} = \overline{aR} = a\overline{R} \cap R$. Therefore $a\overline{R} \subseteq I\overline{R} \subseteq \overline{I}\,\overline{R} \subseteq a\overline{R}$, showing that $I\overline{R} = a\overline{R}$. Conversely, suppose that $I\overline{R} = \alpha\overline{R}$ for some $\alpha \in \overline{R}$. Choose $c \in R$ such that $c\alpha \in R$. Now $\overline{cI} = R \cap (\cap cIV)$ as V ranges over all valuation overrings of R. However $cIV = cI\overline{R}V = c\alpha V$. Thus $\overline{cI} = R \cap (\cap c\alpha V) = R \cap [c\alpha(\cap V)] = R \cap c\alpha\overline{R} = \overline{c\alpha R}$. This shows that $\ell(cI) = 1$. Thus $\ell(I) = 1$.

PROPOSITION 4.6. Let R be a 2-dimensional Noetherian domain with integral closure $\overline{R}$. Let $0 \neq I \subseteq P$ be ideals with P prime. Then $P \in \overline{A}^*(I)$ if and only if one of the following three conditions holds.

(i) P is minimal over I.

(ii) There is a height 1 prime in $\overline{R}$ lying over P.

(iii) $\ell(I_P) > 1$. (Equivalently, $I\overline{R}_S$ is not principal, with $S = R - P$, using Lemma 4.5.)

Proof: If (i) holds, obviously $P \in \overline{A}^*(I)$. If (ii) holds, use Proposition 3.19.

If (iii) holds, since $\dim R = 2$ we have $\ell(I_P) = 2 = \text{height } P$, by Lemma 4.3. Thus Proposition 4.1 gives $P \in \bar{A}^*(I)$.

Conversely, assume $P \in \bar{A}^*(I)$ and that (i) and (iii) both fail to hold. Then clearly height $P = 2$ and $\ell(I_P) = 1$. Proposition 4.1 now shows that R_P does not satisfy the altitude formula (i.e. is not quasi-unmixed). Since R_P is 2-dimensional, the only way this can occur is for $\bar{R}$ to contain a height 1 prime lying over P.

For a 2-dimensional normal Noetherian domain, we find that $\bar{A}^*(I)$ always equals $A^*(I)$.

COROLLARY 4.7. Let R be a normal 2-dimensional Noetherian domain. Let $I \neq 0$ be an ideal. The following are equivalent for a prime P.

(i) $P \in A^*(I)$

(ii) $P \in \bar{A}^*(I)$

(iii) Either P is minimal over I or IR_P is not principal.

Proof: (ii) $\Leftrightarrow$ (iii) is immediate by Proposition 4.6. (ii) $\Rightarrow$ (i) is by Proposition 3.17. Finally, suppose $P \in A^*(I)$, and assume that P is not minimal over I. Then height $P = 2$. For large n, P_P is a prime divisor of I_P^n, which therefore cannot be principal since R_P is a Krull domain. Thus (i) $\Rightarrow$ (iii).

Remarks: (a) The class of domains in which $A^*(I)$ always equals $\bar{A}^*(I)$ will be studied in a later chapter.

(b) The equivalence of (i) and (iii) in Corollary 4.7 was first proved by P. Eakin [ME, Proposition 21]. We use Eakin's arguments in our next result.

PROPOSITION 4.8. Let (R,M) be a 2-dimensional local Cohen-Macaulay domain with multiplicity e. If I is an ideal and $M \in \bar{A}^*(I)$, then $M \in \text{Ass}(R/I^n)$ for all $n \geq e$.

Proof: Let $n \geq e$ and suppose that $M \notin \text{Ass}(R/I^n)$. Then height $I = 1$ and R/I^n is Cohen-Macaulay. Now a result of Rees generalized in [S1] shows that I^n can be generated by e elements. By [ES, Corollary 1] I is prestable, which one easily sees implies $\ell(I) = 1$. As R, being Cohen-Macaulay, satisfies the altitude formula, Proposition 4.1 shows that $M \notin \overline{A}^*(I)$. This contradiction proves the result.

Remark: We do not know if the hypothesis can be weakened to $M \in A^*(I)$ or the conclusion strengthened to $M \in \mathrm{Ass}(R/\overline{I^n})$, $n \geq e$.

CHAPTER V: Asymptotic Sequences

In view of the existence of $A^*(I)$, and the classical concept of an R-sequence, it is natural to consider a sequence of elements $x_1, \ldots, x_n$ such that $(x_1, \ldots, x_n) \neq R$ and for each i, x_i is not in any prime ideal in $A^*((x_1, \ldots, x_{i-1}))$. However, it follows easily from [N1, Exercise 13, p. 103], that such sequences are exactly R-sequences. Instead, we go to $\overrightarrow{A}^*(I)$, and make the following definition.

DEFINITION. The sequence of elements $x_1, \ldots, x_n$ in R is an asymptotic sequence if $(x_1, \ldots, x_n) \neq R$ and if for each $1 \leq i \leq n$, x_i is not in any prime contained in $\overrightarrow{A}^*((x_1, \ldots, x_{i-1}))$.

Remarks: (a) When $i = 1$, we have $x_1 \notin \cup\{P \in \overrightarrow{A}^*(0)\} = \cup\{p \mid p$ is a minimal prime of $R\}$.

(b) If $x_1, \ldots, x_n$ is an asymptotic sequence in R, and if $(x_1, \ldots, x_n) \subseteq P$ with P prime, then $x_1, \ldots, x_n$ is easily seen to be an asymptotic sequence in R_P.

This notion of an asymptotic sequence was independently conceived by Rees, Ratliff, and the author. Rees obtained the first significant result [Rs2], which will be discussed in the next chapter. The main characterization of asymptotic sequences (Proposition 5.4) to be given in this chapter, first appeared in Ratliff's work [R9]. However, Katz independently followed roughly the same trail [Kz1]. We have borrowed extensively from both of these.

LEMMA 5.1. Let (R,M) be a local ring with completion R^*. Let $x_1, \ldots, x_n$ be a sequence of elements in R. The following are equivalent.

(i) $x_1, \ldots, x_n$ is an asymptotic sequence in R.

(ii) $x_1, \ldots, x_n$ is an asymptotic sequence in R^*.

(iii) $x_1 + q, \ldots, x_n + q$ is an asymptotic sequence in R/q for every minimal prime q of R.

Proof: This is an easy consequence of Proposition 3.18.

LEMMA 5.2. Let $x_1, \dots, x_n$ be an asymptotic sequence in R and let P be a prime minimal over $(x_1, \dots, x_n)$. If q is any minimal prime contained in P, then height $P/q = n$.

Proof: We may assume R is local at P. By Lemma 5.1, we may assume that $q = 0$ and R is a domain. We induct on n, the case $n = 1$ being trivial. Now for P minimal over $(x_1, \dots, x_n)$, we obviously have height $P \leq n$. Shrink P to a prime Q minimal over $(x_1, \dots, x_{n-1})$. By induction, height $Q = n - 1$. Since $Q \in \bar{A}^*(x_1, \dots, x_{n-1})$ we have $x_n \in P - Q$. Therefore height $P >$ height $Q = n - 1$. Thus height $P = n$.

Our next result shows that in a quasi-unmixed local ring, $x_1, \dots, x_n$ being an asymptotic sequence is equivalent to $I = (x_1, \dots, x_n)$ being an ideal in the principal class (i.e. height I = minimal number of generators of I). This does not work globally. Let R have two maximal ideals with height $M_1 = 1$ and height $M_2 = 3$. Choose x_1, x_2, x_3 such that height$(x_1, x_2, x_3) = 3$ but with $x_1, x_2 \in M_1$. Since $M_1 \in \bar{A}^*((x_1))$, x_1, x_2, x_3 is not an asymptotic sequence.

LEMMA 5.3. Let (R,M) be a quasi-unmixed local ring. Then $x_1, \dots, x_n$ is an asymptotic sequence if and only if height$(x_1, \dots, x_n) = n$.

Proof: One direction is immediate from Lemma 5.2. Thus suppose that height$(x_1, \dots, x_n) = n$. We induct on n, the case $n = 1$ being trivial. Let Q be a prime minimal over $(x_1, \dots, x_{n-1})$. Therefore height $Q \leq n - 1$, so that $x_n \notin Q$. As R is local, there is a prime P minimal over (Q, x_n). We have height$(P/Q) = 1$, and in view of quasi-unmixedness, height P = height $Q + 1 \leq n$. Since $(x_1, \dots, x_n) \subseteq P$, the hypothesis gives height $P \geq n$. Thus height $P = n$ and so we have height $Q = n - 1$. This shows that height$(x_1, \dots, x_{n-1}) = n - 1$. By induction, $x_1, \dots, x_{n-1}$ is an asymptotic sequence. Now let $p \in \bar{A}^*(x_1, \dots, x_{n-1})$. By Proposition 4.1, height $p = \ell((x_1, \dots, x_{n-1})R_p)$, and of course this analytic spread does not exceed $n - 1$. Therefore height $p \leq n - 1$. By hypothesis, clearly $x_n \notin p$, for any $p \in \bar{A}^*(x_1, \dots, x_{n-1})$. This shows that $x_1, \dots, x_n$ is an asymptotic sequence.

PROPOSITION 5.4. Let (R,M) be a local ring with completion R^*. Let $(x_1,\ldots,x_n) \neq R$. Then $x_1,\ldots,x_n$ is an asymptotic sequence in R if and only if $\text{height}(x_1,\ldots,x_n)R^* + q/q = n$ for every minimal prime q of R^*.

Proof: Since R^*/q is quasi-unmixed, this is immediate from lemmas 5.1 and 5.3.

COROLLARY 5.5. Let $x_1,\ldots,x_n$ be an asymptotic sequence contained in the Jacobson radical of R. Then any permutation of $x_1,\ldots,x_n$ is also an asymptotic sequence.

Proof: It is easily seen that if $x_1,\ldots,x_n$ is in the Jacobson radical, then it is an asymptotic sequence in R if and only if it is an asymptotic sequence in R_M for each maximal ideal M. The result is now easy using Proposition 5.4.

In the following material we will be repeatedly discussing, for a prime P, the minimum of the depths of minimal primes in $(R_P)^*$. We therefore introduce notation.

DEFINITION. If P is prime in a Noetherian ring R, let $z(P) = \min\{\text{depth } q \mid q$ is a minimal prime in $(R_P)^*\}$.

PROPOSITION 5.6. Let I be an ideal in a Noetherian ring R. Let $x_1,\ldots,x_n$ be an asymptotic sequence maximal among asymptotic sequences contained in the ideal I. (That is, there is no $x_{n+1} \in I$ with $x_1,\ldots,x_{n+1}$ an asymptotic sequence.) Then $n = \min\{z(P) \mid P$ is a prime containing $I\} = \min\{z(P) \mid P \in \overline{A}^*(I)\}$.

Proof: Let P be a prime containing I. Now $x_1,\ldots,x_n$ is an asymptotic sequence in R_P. By Proposition 5.4, for any minimal prime q of R_P^*, $\text{height}(x_1,\ldots,x_n)R_P^* + q/q = n$. Thus $\text{depth } q \geq n$, so that $z(P) \geq n$. This gives half of the first equality. Now since $x_1,\ldots,x_n$ is a maximal asymptotic sequence from I, there is a $Q \in \overline{A}^*(x_1,\ldots,x_n)$ with $I \subseteq Q$. Clearly $x_1,\ldots,x_n$ is a maximal asymptotic sequence from Q_Q in R_Q, and by two uses of Proposition 3.18, for some minimal prime q of R_Q^*, $Q_Q^*/q \in \overline{A}^*((x_1,\ldots,x_n)R_Q^* + q/q)$. By

Proposition 4.1, height $Q^*_Q/q = \ell((x_1,\ldots,x_n)R^*_Q + q/q) \le n$. Thus $z(Q) \le n$, proving the first equality.

In order to prove the second equality, it will suffice to show that the Q discussed above is in $\overline{A}^*(I)$, since we clearly have $z(Q) = n$. Therefore we complete the proof of the present proposition by proving the following lemma.

LEMMA 5.7. Let $x_1,\ldots,x_n$ be an asymptotic sequence and let $Q \in \overline{A}^*(x_1,\ldots,x_n)$. If I is an ideal with $(x_1,\ldots,x_n) \subseteq I \subseteq Q$, then $Q \in \overline{A}^*(I)$.

Proof: We may assume that R is local at Q. Obviously $x_1,\ldots,x_n$ is a maximal asymptotic sequence in Q, and so by the first equality in Proposition 5.6 (applied to the ideal Q) $z(Q) = n$. Thus R^* contains a minimal prime p of depth n. By Proposition 5.4, $\text{height}(x_1,\ldots,x_n)R^* + p/p = n$, so that Q^* is minimal over $(x_1,\ldots,x_n)R^* + p$. Thus Q^* is minimal over $IR^* + p$. By Corollary 3.13, $Q^* \in \overline{A}^*(IR^*)$, and by Proposition 3.18, $Q \in \overline{A}^*(I)$.

Remark: The analog of Lemma 5.7 fails for classical R-sequences. That is, if $x_1,\ldots,x_n$ is an R-sequence and if $Q \in \text{Ass}(R/(x_1,\ldots,x_n))$ and if $(x_1,\ldots,x_n) \subseteq I \subseteq Q$, we cannot conclude that $Q \in \text{Ass}(R/I)$.

Proposition 5.6 shows that all asymptotic sequences maximal with respect to coming from I, have the same length.

DEFINITION. Let I be an ideal in a Noetherian ring R. The asymptotic grade of I, $\text{gr}^* I$, is the common length of all maximal asymptotic sequences from I.

We remark that if S is a multiplicatively closed set with $I \cap S = \emptyset$, then $\text{gr}^*(I) \le \text{gr}^*(I_S)$. This is straightforward.

We give several corollaries to Proposition 5.6. The first shows that Cohen-Macaulay rings are to R-sequences as locally quasi-unmixed rings are to asymptotic sequences.

COROLLARY 5.8. The following are equivalent in a Noetherian ring R.

(i) $gr^*(I) = \text{height } I$ for every ideal I of R.

(ii) $gr^*(M) = \text{height } M$ for every maximal ideal M of R.

(iii) R is locally quasi-unmixed.

Proof: (i) $\Rightarrow$ (ii) is trivial. For (ii) $\Rightarrow$ (iii), by Proposition 5.6, we see that $gr^*(M) = z(M)$ for all maximal M. Thus height $M = z(M)$ so that R_M is quasi-unmixed by definition. Thus R is locally quasi-unmixed, and (iii) holds. Finally, to show (iii) $\Rightarrow$ (i), in a locally quasi-unmixed ring, height $P = z(P)$ for all primes P. Therefore by Proposition 5.6 we easily have (iii) $\Rightarrow$ (i).

COROLLARY 5.9. Let (R,M) be a local ring. The following are equivalent.

(i) R is quasi-unmixed.

(ii) There is a system of parameters for R which is an asymptotic sequence.

(iii) Every system of parameters for R is an asymptotic sequence.

Proof: (i) $\Rightarrow$ (iii) is immediate by Lemma 5.3, and (iii) $\Rightarrow$ (ii) is trivial. For (ii) $\Rightarrow$ (i), let $x_1, \ldots, x_n$ be a system of parameters for R which is also an asymptotic sequence. Since $x_1, \ldots, x_n$ is a system of parameters, height $M = n$. Since $x_1, \ldots, x_n$ is an asymptotic sequence, (which is obviously maximal with respect to coming from M), $gr^* M = n$. Since height $M = gr^* M = z(M)$, the last equality by Proposition 5.6, we see that R is quasi-unmixed.

COROLLARY 5.10. Let $x_1, \ldots, x_n$ be an asymptotic sequence and let the prime P be minimal over $(x_1, \ldots, x_n)$. Then R_P is quasi-unmixed.

Proof: Obviously $x_1, \ldots, x_n$ is a maximal asymptotic sequence in P_P. Thus $z(P) = z(P_P) = n$. However, by Lemma 5.2, we also see that height $P = n$. Thus height $P = z(P)$, and we are done.

We now give the promised converse of Proposition 4.1.

COROLLARY 5.11. Let the Noetherian ring R have the property that for any ideal I in the principal class and any prime P, $P \in \bar{A}^*(I)$ implies height $P = \ell(I_P)$. Then R is locally quasi-unmixed.

Proof: Assume that R is not locally quasi-unmixed. Then for some prime Q, R_Q^* contains a minimal prime of depth less than height Q. Thus, using Proposition 5.6, we have $m = gr^* Q \leq z(Q) <$ height Q. Let $x_1, \ldots, x_m$ be a maximal asymptotic sequence coming from Q. By Lemma 5.2, $I = (x_1, \ldots, x_m)$ is in the principal class. Since $x_1, \ldots, x_m$ is a maximal asymptotic sequence coming from Q, there is a prime $P \in \bar{A}^*(x_1, \ldots, x_m)$ with $Q \subseteq P$. Now height $P \geq$ height $Q > m \geq \ell(I_P)$, since analytic spread does not exceed the minimal number of generators. We have now proved the contrapositive of the result.

As another corollary, we get a strengthening of Proposition 3.19 (i) $\Rightarrow$ (ii).

COROLLARY 5.12. Let P be a prime ideal. If I is an ideal contained in P and if $gr^* I = z(P)$, then $P \in \bar{A}^*(I)$.

Proof: Notice that by Proposition 5.6, $z(P) = gr^*(P_P)$. Call this number n. Let $x_1, \ldots, x_n$ be a maximal asymptotic sequence from I. Clearly $x_1, \ldots, x_n$ is also a maximal asymptotic sequence from P_P since $gr^* P_P = z(P_P) = n$. Thus $P_P \in \bar{A}^*((x_1, \ldots, x_n)R_P)$, so that $P \in \bar{A}^*((x_1, \ldots, x_n)R)$. By Lemma 5.7, $P \in \bar{A}^*(I)$.

LEMMA 5.13. Let $x_1, \ldots, x_n$ be an R-sequence from R. Then $x_1, \ldots, x_n$ is an asymptotic sequence. In particular, $gr\ I \leq gr^* I \leq$ height I.

Proof: It follows easily from [K1, Exercise 13, p. 103] that if $P \in A^*(x_1, \ldots, x_i)$ for some i, then $P \in \mathrm{Ass}(R/(x_1, \ldots, x_i))$. Thus, using Proposition 3.17, an R-sequence is easily seen to be an asymptotic sequence. Therefore $gr(I) \leq gr^*(I)$ is obvious. Now for any prime P, clearly $z(P) \leq$ height P, and so by Proposition 5.6, $gr^*(I) \leq$ height I.

Our next result strengthens Proposition 1.10.

PROPOSITION 5.14. Let T be an integral extension of R and let $q_0 \subset q_1 \subset \ldots \subset q_n$ be a saturated chain of primes in T. Suppose that $q_0 \cap R$ is a minimal prime of R. Let $P = q_n \cap R$. Then $\mathrm{gr}\, P \leq \mathrm{gr}^* P \leq n$.

Proof: We already have the first inequality. Now asymptotic upon localization. Thus $\mathrm{gr}^* P \leq \mathrm{gr}^* P_P = z(P)$. We will show that $z(P) \leq n$. We may assume that R is local at P. Letting $p = q_0 \cap R$, $R/p \subseteq T/q_0$ is an integral extension, and T/q_0 contains a maximal chain of length n. Thus, as discussed in the Appendix, $(R/p)^*$ contains a minimal prime of depth n. Now $(R/p)^* \approx R^*/pR^*$, and minimal primes in R^*/pR^* have form p^*/pR^* with p^* a minimal prime of R^*. Thus R^* contains a minimal prime of depth n, so that $z(P) \leq n$.

PROPOSITION 5.15. Let $P \subseteq Q$ be prime ideals in a Noetherian ring. Then $z(Q) \leq z(P) + z(Q/P)$.

Proof: $(R_P)^*$ contains a minimal prime of depth $z(P)$, and $(R_Q/P_Q)^*$ contains a minimal prime of depth $z(Q/P)$. Therefore, a result proved in the Appendix shows that $(R_Q)^*$ contains a minimal prime of depth $z(P) + z(Q/P)$. Thus $z(Q) \leq z(P) + z(Q/P)$.

PROPOSITION 5.16. Let I be an ideal of R and let a be an element of the Jacobson radical of R. Then $\mathrm{gr}^* I \leq \mathrm{gr}^* (I,a) \leq \mathrm{gr}^* I + 1$.

Proof: The first inequality is obvious from the definition of asymptotic grade. We now use Proposition 5.6 to find a prime P containing I, with $\mathrm{gr}^* I = z(P)$. If $(I,a) \subseteq P$ then $\mathrm{gr}^* I \leq \mathrm{gr}^* (I,a) \leq z(P) = \mathrm{gr}^* I$, and we are done. If $(I,a) \not\subseteq P$ then $a \notin P$, but since a is in the Jacobson radical, there is a prime Q minimal over (P,a). Thus height $Q/P = 1$. Obviously $z(Q/P) = 1$, and by Proposition 5.15 $z(Q) \leq z(P) + 1$. Therefore Proposition 5.6 shows that $\mathrm{gr}^* (I,a) \leq z(Q) \leq z(P) + 1 = \mathrm{gr}^* I + 1$.

In Proposition 5.16, the condition that a be in the Jacobson radical is needed. Let R be a domain satisfying the altitude formula and having two maximal ideals M and N, with height $M = 1$ and height $N = 3$. Let $I = M \cap N$ and $a \in N - M$.

Since $z(M) = 1$ and $z(N) = 3$, clearly $gr^*(I) = 1$ while $gr^*(I,a) = 3$. This example also shows that in the next corollary, the local condition is needed.

COROLLARY 5.17. Let I be an ideal in a local ring (R,M), and suppose that $gr^* I < gr^* M$. Then there is a prime P containing I with $gr^* P = gr^* I + 1$.

Proof: Let $x_1, \ldots, x_n$ be a maximal asymptotic sequence from I. Since $gr^* M > n$, there is an $x_{n+1} \in M$ with $x_1, \ldots, x_{n+1}$ an asymptotic sequence. By Proposition 5.16, $gr^*(I,x_{n+1}) = n+1$. Thus $x_1, \ldots, x_{n+1}$ is a maximal asymptotic sequence from (I,x_{n+1}) and so there is a $P \in \overline{A}^*(x_1, \ldots, x_{n+1})$ with $(I,x_{n+1}) \subseteq P$. P is as desired.

PROPOSITION 5.18. Let $I = (a_1, \ldots, a_n)$. Then $gr^*(I) \leq n$. If $gr^* I = n$, then I can be generated by an asymptotic sequence (necessarily of length n).

Proof: By Lemma 5.13, $gr^* I \leq$ height I, which clearly cannot exceed n. The proof of the rest of the result is analogous to the proof of [Kl, Theorem 125].

The example [Kl, Example 7, p. 102] shows that in the above proposition, if $gr^* I = n$, not every minimal basis of I is an asymptotic sequence. However, if I is in the Jacobson radical, it is true.

PROPOSITION 5.19. Let $I = (a_1, \ldots, a_n)$ be in the Jacobson radical of R. Then $a_1, \ldots, a_n$ is an asymptotic sequence if and only if $gr^*(I) = n$.

Proof: One direction is obvious. Suppose $gr^* I = n$. In order to show that $a_1, \ldots, a_n$ is an asymptotic sequence in R, we easily see that it is enough to show they form an asymptotic sequence in R_M for each maximal ideal M. Now $n = gr^* I \leq gr^* I_M \leq n$ by Proposition 5.18. Thus we may assume R local at M. Proposition 5.18 shows that $I = (b_1, \ldots, b_n)$ with $b_1, \ldots, b_n$ an asymptotic sequence. Now by Proposition 5.4, for any minimal prime q of R^*, $\text{height}(b_1, \ldots, b_n)R^* + q/q = n$. Thus $\text{height}(a_1, \ldots, a_n)R^* + q/q = n$ and Proposition 5.4 shows that $a_1, \ldots, a_n$ is an asymptotic sequence.

COROLLARY 5.20. Let $I = (a_1, \ldots, a_n)$ and $J = (b_1, \ldots, b_n)$ be two ideals in the Jacobson radical of R. Suppose Rad I = Rad J. Then $a_1, \ldots, a_n$ is an asymptotic sequence if and only if $b_1, \ldots, b_n$ is an asymptotic sequence.

Proof: Proposition 5.6 shows that $\mathrm{gr}^* I = \mathrm{gr}^* J$. The result is now immediate from the preceeding proposition.

The analogue of the unmixedness theorem for R-sequences holds.

PROPOSITION 5.21. Let $\mathrm{gr}^* I = n$ and suppose that I can be generated by n elements. If $P \in \mathrm{Ass}(R/\overline{I^m})$ for any m, then $\mathrm{gr}^* P = n$.

Proof: By Proposition 5.18 we may assume that $I = (a_1, \ldots, a_n)$ with $a_1, \ldots, a_n$ an asymptotic sequence. If $P \in \bar{A}(I,m)$ then by Proposition 3.9, $P \in \bar{A}^*(a_1, \ldots, a_n)$, so that $a_1, \ldots, a_n$ is a maximal asymptotic sequence coming from P. Thus $\mathrm{gr}^* P = n$.

We show two ways asymptotic sequences differ from R-sequences.

If $x_1, \ldots, x_n$ is an R-sequence in R and if the images of $x_{n+1}, \ldots, x_m$ are an R-sequence in $R/(x_1, \ldots, x_n)$, then $x_1, \ldots, x_n, x_{n+1}, \ldots, x_m$ is an R-sequence in R. The analogue for asymptotic sequences fails. Let (R,M) be a local ring with exactly two minimal primes p and q, with depth p = 1 and depth q = 2. Pick $a \in M - (q \cup p)$ and $b \in M$ with b not in any prime minimal over (q, a). Since $a \notin q \cup p$, a is an asymptotic sequence in R. Also, the image of b is easily seen to not be in any minimal prime of R/aR. Thus that image is an asymptotic sequence in R/aR. Now a,b is not an asymptotic sequence in R, since depth p = 1 implies z(M) = 1, so that $\mathrm{gr}^*(M) = 1$.

We have previously noted that if $x_1, \ldots, x_n$ is an R-sequence, and if $P \in A^*(x_1, \ldots, x_n)$ then $P \in \mathrm{Ass}(R/(x_1, \ldots, x_n))$. The analogue for asymptotic sequences fails. To see this, as in [N, Example 2, pp. 203-205] let (R,M) be a 2-dimensional local domain with $\bar{R}$ containing exactly two maximal ideals, N_1 and N_2, with heights 1 and 2 respectively, such that $M = N_1 \cap N_2$ and such that $\bar{R} = R + R\alpha$ with $N_1 = \alpha\bar{R}$. Let $\beta \in N_2 - N_1$. Note that $\alpha\beta \in N_1 \cap N_2 = M$. We claim that $\overline{\alpha\beta R} = \beta\overline{R} \cap R$. Clearly $\overline{\alpha\beta R} \subseteq \alpha\beta\overline{R} \cap R \subseteq \beta\overline{R} \cap R$. Now if $r_1 + r_2\alpha \in R + R\alpha = \overline{R}$ with $\beta(r_1 + r_2\alpha) = s \in R$, then to show $s \in \overline{\alpha\beta R}$ it is enough to show $\beta r_1 \in \overline{\alpha\beta R}$ since clearly $\beta r_2\alpha \in \overline{\alpha\beta R}$. Since $\beta \in N_2$, $s \in M$. Also $\beta r_2\alpha \in M$. Thus $\beta r_1 \in M = N_1 \cap N_2$. As $\beta \notin N_1$, $r_1 \in N_1 = \alpha\overline{R}$. Therefore $\beta r_1 \in \beta\alpha\overline{R} \cap R = \overline{\alpha\beta R}$ as desired. Thus $\overline{\alpha\beta R} = \beta\overline{R} \cap R$ as claimed. Now $a = \alpha\beta$ is an asymptotic sequence in R, and since height $N_1 = 1$ implies that $z(M) = 1$ so that $\mathrm{gr}^* M = 1$, we have $M \in \overline{A}^*((a))$. However $M \notin \mathrm{Ass}(R/\overline{(a)})$ since if $M = (\overline{(a)}: b)$ for some $b \in R$, then $\overline{(a)} = \overline{\alpha\beta R} = \beta\overline{R} \cap R$ would imply $M = (\beta\overline{R}: b)_{\overline{R}} \cap R$. By Lemma 1.2 we see that N_2 is a prime divisor of $\beta\overline{R}$. This is impossible since $\overline{R}$ is a Krull domain and height $N_2 = 2$.

The next proposition is due to Keith Whittington.

PROPOSITION 5.22. The following are equivalent for a Noetherian domain R.

i) R satisfies the altitude formula,

ii) for any finitely generated extension domain T of R, and any height 1 prime Q of $\overline{T}$, height $Q \cap T = 1$.

Proof: i) $\Rightarrow$ ii) is trivial since a finitely generated extension domain of a Noetherian domain satisfying the altitude formula also satisfies it.

ii) $\Rightarrow$ i): Suppose that i) fails. Then by Proposition 5.11, there is an ideal $I = (a_1, \ldots, a_n)$ of the principal class and a prime $P \in \overline{A}^*(I)$ with height $P > \ell(I_P)$. Note that I_P is in the principal class in R_P so that $\ell(I_P) = n$. By Proposition 3.20, we may assume that $T = R[a_1/a_n, \ldots, a_{n-1}/a_n]$ is such that $\overline{T}$ contains a height 1 prime Q lying over P. By [D, Corollary 2], we see that PT is a prime of height equal to height $P - (n-1) > 1$. Since $PT \subseteq Q \cap T$, (ii) fails.

CHAPTER VI: Asymptotic Sequences Over Ideals

The first published result involving asymptotic sequences was by Rees [Rs2]. He defined an asymptotic sequence over an ideal, as follows. (Actually, he called it an asymptotic prime sequence over I.)

DEFINITION. Let I be an ideal in a Noetherian ring R. Elements $x_1, \ldots, x_n$ of R are called an asymptotic sequence over I if $(I, x_1, \ldots, x_n) \neq R$ and if for all $i = 1, \ldots, n$, $x_i \notin \bar{A}^*(I, x_1, \ldots, x_{i-1})$.

Rees then let r be the length of a maximal asymptotic sequence over I in a local ring (R,M), and showed that $r \leq \text{height } M - \ell(I)$, and that equality holds if R is quasi-unmixed.

As we now have more tools at our command than did Rees, we will go a bit further, showing that in any local ring, all maximal asymptotic sequences over I have the same length, which we characterize. This was done by Katz in [Kz2]. We then show that if $x_1, \ldots, x_n$ is an asymptotic sequence over I, it is also an asymptotic sequence modulo I.

We begin with a nice result of Rees, proved in [Rs2, Theorem 2.6].

PROPOSITION 6.1. Let I be an ideal in a local ring (R,M). Let $\mathcal{R} = R[t^{-1}, It]$ be the Rees ring of R with respect to I, and let $x \in M$ with x not in any minimal prime divisor of $t^{-1}\mathcal{R}$. Then $\ell(I,x) = \ell(I) + 1$.

Proof: Write $\mathcal{R}(I)$ and $\mathcal{R}(I,x)$ for the Rees rings of R with respect to I and (I,x) respectively. Also let Y be an indeterminate and write $\mathcal{R}(I,Y)$ for the Rees ring of $R[Y]$ with respect to (I,Y). Let $\varphi: R[Y] \to R$ via $Y \to x$. This induces an obvious homomorphism $\tilde{\varphi}$ of $\mathcal{R}(I,Y)$ onto $\mathcal{R}(I,x)$. We claim that $\ker \tilde{\varphi} \subseteq \text{rad}[(t^{-1},x)\mathcal{R}(I,Y)]$. However, we will defer the proof of the claim briefly.

Using the above claim, since $x \in M$, we see that $\ker \tilde{\varphi} \subseteq \text{rad}[(t^{-1},M,Y)\mathcal{R}(I,Y)]$. Therefore, since $\tilde{\varphi}$ maps $(t^{-1},M,Y)\mathcal{R}(I,Y)$ onto $(t^{-1},M)\mathcal{R}(I,x)$, we see that $\tilde{\varphi}$ induces an isomorphism from $\mathcal{R}(I,Y)/\sqrt{(t^{-1},M,Y)\mathcal{R}(I,Y)}$ onto $\mathcal{R}(I,x)/\sqrt{(t^{-1},M)\mathcal{R}(I,x)}$. Furthermore, $\tilde{\varphi}$ carries the maximal ideal $N = \ldots + R[Y]t^{-1} + (M,Y) + (I,Y)t + \ldots$

of $\mathcal{R}(I,Y)$ to the maximal ideal $N' = \ldots + Rt^{-1} + M + (I,x)t + \ldots$ of $\mathcal{R}(I,x)$. Since $\ell(I,x)$ is the height of $N'/(t^{-1},M)\mathcal{R}(I,x)$, our isomorphism shows that $\ell(I,x)$ equals the height of $N/(t^{-1},M,Y)\mathcal{R}(I,Y)$.

Since $Y = t^{-1}(Yt)$, we see that $\mathcal{R}(I,Y) = R[Y,t^{-1},(I,Y)t] = R[t^{-1},It,Yt] = \mathcal{R}(I)[Yt]$. Letting $Z = Yt$, Z is an indeterminate over $\mathcal{R}(I)$ and we have $\mathcal{R}(I,Y) = \mathcal{R}(I)[Z]$. Now $N = (t^{-1},M,Y,(I,Y)t)\mathcal{R}(I,Y)$ when written in terms of Z becomes $(t^{-1},M,Zt^{-1},It,Z)\mathcal{R}(I)[Z] = (p,Z)\mathcal{R}(I)[Z]$ with $p = (t^{-1},M,It)\mathcal{R}(I) = \ldots + Rt^{-1} + M + It + \ldots$. Similarly, $(t^{-1},M,Y)\mathcal{R}(I,Y)$ is $q\mathcal{R}(I)[Z]$ with $q = (t^{-1},M)\mathcal{R}(I) = \ldots Rt^{-1} + M + MIt + \ldots$. Thus $\ell(I,x) = \text{height } N/(t^{-1},M,Y)\mathcal{R}(I,Y) = \text{height } (p,Z)\mathcal{R}(I)[Z]/q\mathcal{R}(I)[Z] = \text{height } p/q + 1 = \ell(I) + 1$.

We now return to proving our earlier claim. First note that $\ker \varphi = (Y-x)R[Y]$ shows that $\ker \tilde{\varphi} = \Sigma[(Y-x) \cap (I,Y)^n]t^n$, $n \in Z$. Since for $m \leq 0$, $(I,Y)^m = R[Y]$, we see that if $\alpha \in \ker \tilde{\varphi}$, then for large n, $\alpha t^{-n} \in (Y-x)\mathcal{R}(I,Y)$. Let $\mathcal{P}$ be minimal over $(t^{-1},x)\mathcal{R}(I,Y) = (t^{-1},Y-x)\mathcal{R}(I,Y)$ (since $Y = t^{-1}Z$), and assuming our claim to be false, choose $\alpha \in \ker \tilde{\varphi} - \mathcal{P}$. For large n, $\alpha t^{-n} \in (Y-x)\mathcal{R}(I,Y)$, and since $\alpha \notin \mathcal{P}$ we see that after localizing at $\mathcal{P}$, that t^{-1} is in the radical of the ideal generated by $Y - x$. Since $\mathcal{P}$ is minimal over $(t^{-1},Y-x)\mathcal{R}(I,Y)$, we now see that $\mathcal{P}$ is minimal over $(Y-x)\mathcal{R}(I,Y)$. Thus height $\mathcal{P} \leq 1$. Thinking of $\mathcal{P}$ as a prime in $\mathcal{R}(I)[Z]$, we must have height $\mathcal{P} \cap \mathcal{R}(I) \leq 1$. As t^{-1} is a regular element in $\mathcal{P} \cap \mathcal{R}(I)$, height $\mathcal{P} \cap \mathcal{R}(I) = 1$. By the choice of x, $x \notin \mathcal{P} \cap \mathcal{R}(I)$, a contradiction since $x \in \mathcal{P}$. This completes the proof of the claim, and of the proposition.

COROLLARY 6.2. If $x_1, \ldots, x_r$ is an asymptotic sequence over I in a local ring, then $\ell(I,x_1,\ldots,x_r) = \ell(I) + r$.

Proof: This follows from Proposition 6.1 and Proposition 3.10.

DEFINITION. $x_1, \ldots, x_r$ is a maximal asymptotic sequence over I if it is an asymptotic sequence over I but $x_1, \ldots, x_r, x_{r+1}$ is not an asymptotic sequence over I, for any x_{r+1}.

We now show that in a local ring, all maximal asymptotic sequences over I have the same length.

PROPOSITION 6.3. [Kz2] Let I be an ideal in the local ring (R,M). Let $x_1,\dots,x_r$ be a maximal asymptotic sequence over I. Then $r = \min\{\text{height}(M^*/q) - \ell(IR^*+q/q) \mid q \text{ is a minimal prime in the completion } R^*\}$.

Proof: It follows easily from Proposition 3.18 that for any minimal prime q of R^*, $x_1+q,\dots,x_r+q$ is an asymptotic sequence over IR^*+q/q. By Lemma 4.3 and Corollary 6.2, $\ell(IR^*+q/q)+r = \ell((I,x_1,\dots,x_r)R^*+q/q) \leq \text{height } M^*/q$. Now by the maximality of $x_1,\dots,x_r$, $M \in \bar{A}^*(I,x_1,\dots,x_r)$ and so Proposition 3.18 shows that for some minimal prime q of R^*, $M^*/q \in \bar{A}^*((I,x_1,\dots,x_r)R^*+q/q)$. Since R^*/q satisfies the altitude formula, by Proposition 4.1, height $M^*/q = \ell((I,x_1,\dots,x_r)R^*+q/q) = \ell(IR^*+q/q)+r$, and we are done.

COROLLARY 6.4. Let r be the length of a maximal asymptotic sequence over I in a local ring (R,M). Then

(i) $r \leq \text{height } M - \ell(I)$

(ii) The following are equivalent.

(a) R is quasi-unmixed

(b) Equality holds in (i) for all ideals I

(c) Equality holds in (i) for all ideals I in the principal class.

Proof: (i) By Lemma 4.2, there is a minimal prime q of R^* with $\ell(I) = \ell(IR^*) = \ell(IR^*+q/q)$. By Proposition 6.3, $r \leq \text{height}(M^*/q) - \ell(IR^*+q/q) = \text{height}(M^*/q) - \ell(I) \leq \text{height } M - \ell(I)$.

(ii) (a) $\Rightarrow$ (b). This is easy using Proposition 6.3, Lemma 4.2, and the fact that height $M = \text{height}(M^*/q)$ for any minimal prime q in the completion of a quasi-unmixed local ring.

(b) $\Rightarrow$ (c) is immediate.

(c) $\Rightarrow$ (a). Let $gr^*M = z(M) = n$, and let $x_1,\dots,x_n$ be a maximal asymptotic sequence from M. Let $I = (x_1,\dots,x_n)$. Now Lemma 5.2 shows that I is in the principal class. The maximality of our sequence gives $M \in \bar{A}^*(I)$, so in this case, $r = 0$. By (c), height $M = \ell(I)$. But $\ell(I)$ lies between height I and the number of generators of I, both of which are n. Thus $\ell(I) = n$. Therefore height $M = n = z(M)$, showing that R is quasi-unmixed.

We now wish to show that if $x_1, \ldots, x_n$ is an asymptotic sequence over I, then it is also an asymptotic sequence modulo I (but the converse fails). We start by showing the numbers compare correctly (Proposition 6.6). First, a lemma.

LEMMA 6.5. Let I be an ideal in a local ring (R,M) and let P be a prime minimal over I. Then $\ell(I) \geq$ height P.

Proof: Since the minimal number of generators of I^n is at least as great as the minimal number of generators of I_P^n for all n, the Hilbert polynomial characterization of analytic spread shows that $\ell(I) \geq \ell(I_P)$. Now I_P is P_P-primary, so it is well known that $\ell(I_P) =$ height P.

PROPOSITION 6.6. Let I be an ideal in a local ring (R,M) and let r be the length of a maximal asymptotic sequence over I. Then $r \leq gr^*(M/I)$.

Proof: Since $(R/I)^* \approx R^*/IR^*$, there is a prime p of R^*, minimal over IR^*, such that $gr^*(M/I) = \text{depth } p$. Let q be a minimal prime of R^* with $q \subseteq p$. Since R^*/q is quasi-unmixed, $gr^*(M/I) = \text{depth } p = \text{height}(M^*/q) - \text{height}(p/q) \geq \text{height}(M^*/q) - \ell(IR^*+q/q)$ (by Lemma 6.5) $\geq r$ (by Proposition 6.3).

LEMMA 6.7. Let I be an ideal and let $x_1, \ldots, x_n$ be an asymptotic sequence over I and also an asymptotic sequence modulo I. Let P be a prime containing I such that $P/I \in \bar{A}^*(x_1+I, \ldots, x_n+I)$. Then $P \in \bar{A}^*(I, x_1, \ldots, x_n)$.

Proof: We may localize, and assume that R is local at P. The hypothesis shows that $x_1+I, \ldots, x_n+I$ is a maximal asymptotic sequence from P/I. Thus $gr^*(P/I) = n$. By Proposition 6.6, in the ring (R,P) the length of a maximal asymptotic sequence over I cannot exceed n. As $x_1, \ldots, x_n$ is an asymptotic sequence over I, it must be maximal. Thus $P \in \bar{A}^*(I, x_1, \ldots, x_n)$.

PROPOSITION 6.8. Let $x_1, \ldots, x_n$ be an asymptotic sequence over I. Then $x_1+I, \ldots, x_n+I$ is an asymptotic sequence in R/I.

Proof: We induct on n. For $n=1$, the hypothesis is that x_1 is not in any prime contained in $\overrightarrow{A}^*(I)$. In particular, x_1 is not in any prime minimal over I. Thus x_1+I is an asymptotic sequence in R/I. Suppose now that we have $x_1+I,\ldots,x_i+I$ an asymptotic sequence in R/I for $i<n$. Let P/I be prime in R/I with $P/I \in \overrightarrow{A}^*(x_1+I,\ldots,x_i+I)$. Then Lemma 6.7 shows that $P \in \overrightarrow{A}^*(I,x_1,\ldots,x_i)$. Since $x_1,\ldots,x_i,x_{i+1}$ is an asymptotic sequence over I, $x_{i+1} \notin P$. Thus $x_{i+1}+I \notin P/I$, and we see that $x_1+I,\ldots,x_{i+1}+I$ is an asymptotic sequence in R/I.

The converse of Proposition 6.8 is easily seen to fail, since $\overrightarrow{A}^*(I)$ can contain primes not minimal over I.

PROPOSITION 6.9. Let I be an ideal in a local ring (R,M), and let $y_1,\ldots,y_n$ be an asymptotic sequence from I. Then there is a maximal asymptotic sequence over I, $x_1,\ldots,x_r$, such that $y_1,\ldots,y_n,x_1,\ldots,x_r$ is an asymptotic sequence. In particular, $r+\mathrm{gr}^*I \leq \mathrm{gr}^*M$.

Proof: Let r be the length of a maximal asymptotic sequence over I. If $r=0$, we are done. If $r>0$, then $M \notin \overrightarrow{A}^*(I)$, and so by Lemma 5.7, $M \notin \overrightarrow{A}^*(y_1,\ldots,y_n)$. Pick $x_1 \in M$ with $x_1 \notin \cup\{P \in \overrightarrow{A}^*(I)\}$ and $x_1 \notin \cup\{P \in \overrightarrow{A}^*(y_1,\ldots,y_n)\}$. Now x_1 is an asymptotic sequence over I, and the length of a maximal asymptotic sequence over (I,x_1) is $r-1$ (since if $x_1,x_2',\ldots,x_r'$ is a maximal asymptotic sequence over I, then $x_2',\ldots,x_r'$ is a maximal asymptotic sequence over (I,x_1)). Since the choice of x_1 assures that $y_1,\ldots,y_n,x_1$ is an asymptotic sequence, we may use induction.

We now determine when the inequality in Proposition 6.9 is equality. First we need another characterization of gr^*I.

PROPOSITION 6.10. Let I be an ideal in a local ring (R,M). Then $\mathrm{gr}^*(I) = \min\{\text{height } IR^*+q/q \mid q \text{ is a minimal prime in } R^*, \text{ the completion of } R\}$.

Proof: Let $x_1,\ldots,x_n$ be a maximal asymptotic sequence from I. By Lemma 5.1, $x_1+q,\ldots,x_n+q$ is an asymptotic sequence in R^*/q for any minimal prime q of R^*.

Thus $gr^*(IR^*+q/q) \geq n$. As R^*/q is quasi-unmixed, Corollary 5.8 gives height $IR^*+q/q \geq n = gr^*I$. Now pick $P \in \overline{A}^*(x_1,\ldots,x_n)$ with $I \subseteq P$. By two uses of Proposition 3.18, there is a P^* of R^* with $P^* \cap R = P$ and a minimal prime $q \subseteq P^*$, with $P^*/q \in \overline{A}^*((x_1,\ldots,x_n)R^*+q/q)$. Corollary 5.8 shows that height $P^*/q = gr^*P^*/q = n$. As $P^* \cap R = P$, $IR^*+q/q \subseteq P^*/q$, so height $IR^*+q/q \leq n = gr^*I$. As we already have the other inequality, we are done.

PROPOSITION 6.11. Let I be an ideal in a local ring (R,M), and let r be the length of a maximal asymptotic sequence over I. The following are equivalent.

(i) $r + gr^*I = gr^*M$.

(ii) For some minimal prime q of R^*, $\ell(IR^*+q/q) = \text{height}(IR^*+q/q) = gr^*I$ and $r = \text{depth } q - \text{height}(IR^*+q/q)$.

(iii) The equalities in (ii) hold for every minimal prime q of R^* satisfying depth $q = gr^*M$.

Proof: (i) $\Rightarrow$ (iii): Let q be any minimal prime of R^* with depth $q = gr^*M$. We have that depth q = height M^*/q = (height M^*/q - height(IR^*+q/q)) + height(IR^*+q/q) $\geq$ (height $M^*/q - \ell(IR^*+q/q)$) + height(IR^*+q/q), the inequality holding since height does not exceed analytic spread. Now using Propositions 6.3 and 6.10, (height $M^*/q - \ell(IR^*+q/q)$) + height$(IR^*+q/q) \geq r + gr^*I$. Using our choice of q, and (i), we also have $r + gr^*I = gr^*M$ = depth q. As we began and ended with depth q, we see that equality holds throughout, and (iii) easily follows.

(iii) $\Rightarrow$ (ii) is immediate.

(ii) $\Rightarrow$ (i). Let q be as in (ii). We have that $gr^*M \leq$ depth q = height M^*/q = (height M^*/q - height(IR^*+q/q)) + height$(IR^*+q/q) = r + gr^*I \leq gr^*M$, the last two steps using (ii) and Proposition 6.9. Thus (i) holds.

COROLLARY 6.12. Let I be an ideal in a local ring (R,M), and let r be the length of a maximal asymptotic sequence over I. Then $r + gr^*I$ = height M if and only if R is quasi-unmixed and height $I = \ell(I)$.

Proof: Assume that $r + gr^* I = $ height M. Since $r + gr^* I \leq gr^* M \leq $ height M, we have $gr^* M = $ height M, which shows that R is quasi-unmixed. By Corollary 6.4, $r + \ell(I) = $ height M. The hypothesis and Corollary 5.8 show $\ell(I) = gr^* I = $ height I. Conversely, if R is quasi-unmixed and height $I = \ell(I)$, then Corollaries 6.4 and 5.8 give $r + gr^* I = $ height M.

Our next lemma is due to Ratliff.

LEMMA 6.13. Let $P \in \bar{A}^*(I)$, let $a_1, \ldots, a_n$ be an asymptotic sequence over I, and let Q be a prime minimal over $(P, a_1, \ldots, a_n)$. Then $Q \in \bar{A}^*(I, a_1, \ldots, a_n)$.

Proof: We may localize at Q, and then using Proposition 3.18, we may go to the completion and the work modulo a minimal prime. That is, we may assume that (R, Q) is a complete domain, which must be quasi-unmixed. Now as mentioned in the proof of Lemma 6.5, $\ell(I) \geq \ell(I_p)$. Using Corollary 6.2 and Proposition 4.1,

$\ell(I, a_1, \ldots, a_n) = \ell(I) + n \geq \ell(I_p) + n \geq \ell(I_p) + n = $ height $P + n \geq $ height $P + $ height $Q/P = $ height Q since R is catenary. Using Lemma 4.3, we see that $\ell(I, a_1, \ldots, a_n) = $ height Q. Thus Proposition 4.1 gives the result.

In Lemma 6.13 it is not enough to assume that $Q \in \bar{A}^*(P, a_1, \ldots, a_n)$. To see this, consider the example and remark preceeding Lemma 4.5. That example gave a 3-dimensional quasi-unmixed local ring (R, M) and a height 1 prime P with $\ell(P) = 2$, such that $\bar{A}^*(P) = \{P\}$. Pick $0 \neq x \in P$ and let $I = xR$. Then $\ell(I) = 1$. By Proposition 4.1, $M \notin \bar{A}^*(I)$. Let a be an asymptotic sequence over I. Thus $a \notin P$. Since $\bar{A}^*(P) = \{P\}$, a is also an asymptotic sequence over P. By Corollary 6.2, $\ell(I, a) = \ell(I) + 1 = 2$ while $\ell(P, a) = \ell(P) + 1 = 3$. By Proposition 4.1, $M \in \bar{A}^*(P, a)$ but $M \notin \bar{A}^*(I, a)$.

PROPOSITION 6.14. Let I be an ideal in a local ring (R, M), and let r be the length of a maximal asymptotic sequence over I. Then $r \leq \min\{$little depth $P \mid P \in \bar{A}^*(I)\}$ (little depth P is the length of a shortest saturated chain of primes between P and M.)

Proof: Let $P \in \overline{A}^*(I)$ and suppose that $n = $ little depth P is minimal among primes in $\overline{A}^*(I)$. Consider a chain of primes $P = P_0 \subset P_1 \subset \ldots \subset P_{n-1} \subset P_n = M$. We induct on n. If $n = 0$, $M = P \in \overline{A}^*(I)$ and so $r = 0$. If $n = 1$, then by definition of n, $M \notin \overline{A}^*(I)$ and we pick $a \in M$ to be an asymptotic sequence over I. Since $a \notin P$ and since $P \subset M$ is saturated, M is minimal over (P,a). By Lemma 6.13, $M \in \overline{A}^*(I,a)$ showing that $r = 1$. Now assume $n > 1$. By [M2, Theorem 5] we can assume that height $\frac{P_{n-1}}{P_0} = n-1$. Now the definition of n, and the fact that $n > 1$, show that neither P_{n-1} nor $M = P_n$ are in $\overline{A}^*(I)$. Thus $P_{n-1} \not\subset \cup\{Q \in \overline{A}^*(I)\}$. Pick $a \in P_{n-1}$ with a an asymptotic sequence over I. Of course $a \notin P$. Let $(P,a) \subset P_1' \subset P_{n-1}$ with P_1' a prime minimal over (P,a). Thus height $\frac{P_1'}{P_0} = 1$, and since height $\frac{P_{n-1}}{P_0} = n-1$ we must have height $\frac{P_{n-1}}{P_1'} \leq n-2$. Thus since $P_{n-1} \subset M$ is saturated, we see that little depth $P_1' \leq n-1$. By Lemma 6.13, we see that $P_1' \in \overline{A}^*(I,a)$. As the length of a maximal asymptotic sequence over (I,a) is $r-1$, by induction we have $r-1 \leq n-1$. Thus $r \leq n$.

In a quasi-unmixed local ring, we can characterize when the inequality of Proposition 6.14 is equality. Of course since R is catenary, little depth equals depth.

PROPOSITION 6.15. Let I be an ideal in a quasi-unmixed local ring (R,M) and let r be the length of a maximal asymptotic sequence over I. The following are equivalent.

i) $r = \min\{\text{depth } P \mid P \in \overline{A}^*(I)\}$

ii) $\ell(I) = \max\{\text{height } P \mid P \in \overline{A}^*(I)\}$

iii) $\ell(I) = \ell(I_P)$ for some $P \in \overline{A}^*(I)$.

Proof: By Proposition 6.4, $r + \ell(I) = $ height M. Since R is catenary, depth P + height $P = $ height M for all primes P. The equivalence of i) and ii) is now straightforward. ii) $\Rightarrow$ iii) is immediate from Proposition 4.1. Finally, suppose $\ell(I) = \ell(I_P)$ with $P \in \overline{A}^*(I)$. By Proposition 4.1, $\ell(I) = $ height P. The truth of ii) now follows from the next lemma.

LEMMA 6.16. Let I be an ideal in a quasi-unmixed local ring. Then $\ell(I) \geq$ height P for all $P \in \overrightarrow{A}^*(I)$.

Proof: $\ell(I) \geq \ell(I_P)$ as we have already seen. Now use Proposition 4.1.

PROPOSITION 6.17. Let I be an ideal in a local ring (R,M) and let $x_1, \ldots, x_n$ be an asymptotic sequence over I. Then any permutation of $x_1, \ldots, x_n$ is also an asymptotic sequence over I.

Proof: Since $x_1, \ldots, x_n$ is an asymptotic sequence over I if and only if $x_1 + q, \ldots, x_n + q$ is an asymptotic sequence over $IR^* + q/q$ for any minimal prime q of R^* (Proposition 3.18), we may assume that R is quasi-unmixed. Also note that $x_1, \ldots, x_n$ is an asymptotic sequence over I if and only if $x_1, \ldots, x_i$ is an asymptotic sequence over I, and $x_{i+1}, \ldots, x_n$ is an asymptotic sequence over $(I, x_1, \ldots, x_i)$. This, together with the fact that any permutation is a product of adjacent transpositions, easily allows us to reduce to the following. We assume that x,y is an asymptotic sequence over I, and must show that y,x is an asymptotic sequence over I.

Suppose that $P \in \overrightarrow{A}^*(I)$ and $y \in P$. Let Q be a prime minimal over (P,x). By Lemma 6.13, $Q \in \overrightarrow{A}^*(I,x)$. As $y \in P \subseteq Q$, our assumption on x,y is contradicted. Therefore we see that y is an asymptotic sequence over I.

Now suppose $p \in \overrightarrow{A}^*(I,y)$ and $x \in p$. We may localize at p, maintaining all of our hypotheses. By Proposition 4.1 and Corollary 6.2, height $p = \ell(I,y) = \ell(I) + 1$. However, Lemma 4.3 shows that height $p \geq \ell(I,x,y) = \ell(I) + 2$. This contradiction completes the proof.

LEMMA 6.18. Let I be an ideal in a local ring R, and let I' be the image of I in a homomorphic image of R. Then $\ell(I') \leq \ell(I)$.

Proof: For all n, I^n requires at least as many generators as I'^n. Thus the lemma follows from the Hilbert polynomial characterization of analytic spread.

LEMMA 6.19. Let I be an ideal in a locally quasi-unmixed Noetherian ring R. Let $y_1, \ldots, y_m$ be elements of R which form both an asymptotic sequence and an asymptotic sequence over I. Let primes denote modulo $(y_1, \ldots, y_m)$. If $P' \in \overrightarrow{A}^*(I')$, then $P \in \overrightarrow{A}^*(I, y_1, \ldots, y_m)$.

Proof: We localize at P. Lemma 5.2 implies that R' is quasi-unmixed. Proposition 4.1 says height $P' = \ell(I')$. Now using the principal ideal theorem, Lemma 6.18, and Corollary 6.2, we have height $P \leq$ height $P' + m = \ell(I') + m \leq \ell(I) + m = \ell(I, y_1, \ldots, y_m)$. Finally Lemma 4.3 shows height $P = \ell(I, y_1, \ldots, y_m)$ and so Proposition 4.1 gives the result.

PROPOSITION 6.20. Let I be an ideal in a local ring (R,M), and let $x_1, \ldots, x_n$ be an asymptotic sequence over I. Then $x_1, \ldots, x_n$ is an asymptotic sequence.

Proof: By Proposition 5.1, it is enough to show that the images of $x_1, \ldots, x_n$ form an asymptotic sequence in R^*/q for each minimal prime q of R^*. As Proposition 3.18 implies that those images are an asymptotic sequence over IR^*+q/q, we may return to (R,M) with the added assumption that it is a quasi-unmixed local ring.

Suppose x_1 is in q, a minimal prime of R. Choose P minimal over $I+q$. By Corollary 3.13, $P \in \overrightarrow{A}^*(I)$. As $x_1 \in P$, we have a contradiction. Thus x_1 is not in any minimal prime, and so is an asymptotic sequence. Now inductively assume that $x_1, \ldots, x_{n-1}$ is an asymptotic sequence. Letting primes denote modulo $(x_1, \ldots, x_{n-1})$, we claim that x_n' is an asymptotic sequence over I'. This is immediate, using Lemma 6.19. Theorefore, the case $n = 1$ shows that x_n' is an asymptotic sequence in R'. Thus x_n is not in any prime minimal over $(x_1, \ldots, x_{n-1})$. By Lemma 5.3, height$(x_1, \ldots, x_{n-1}) = n-1$, so that height $(x_1, \ldots, x_n) = n$. Lemma 5.3 now gives the result.

We wish to show that if $x_1, \ldots, x_n$ is an asymptotic sequence over I in a local ring, then $\ell(I) = \ell((I, x_1, \ldots, x_n)/(x_1, \ldots, x_n))$. We begin with a lemma due to Brodmann, whose hypothesis will be significant in the next chapter.

LEMMA 6.21. Let I be an ideal in a local ring (R,M) and suppose the images of

$x_1, \ldots, x_n$ modulo I^m are an R-sequence in R/I^m for all large m. Then $\ell(I) = \ell((I, x_1, \ldots, x_n)/(x_1, \ldots, x_n))$.

Proof: We induct on n. For $n = 1$, we have $x_1 \notin \cup\{P \in A^*(I)\}$. We claim that for large n, $I^m/MI^m \approx I'^m/M'I'^m$, the prime denoting modulo x_1. To see this, note that if $y \in I^m$ and $y' \in M'I'^m$, then $y = z + rx_1$ with $z \in MI^m$. As $rx_1 = y - z \in I^m$ the hypothesis on x_1 shows that $r \in I^m$, so that $rx \in MI^m$. Thus $y \in MI^m$. Now this isomorphism shows that I^m and I'^m require an equal number of generators. Thus $\ell(I) = \ell(I')$, proving the case $n = 1$.

We inductively assume the result for $n - 1$, and prove it for n. It is straight forward to verify that $R/I^m/x_1(R/I^m) \approx R'/I'^m$. Since the images of $x_2, \ldots, x_n$ are an R-sequence on the left hand side, going to the right hand side and applying the case $n - 1$, we have $\ell(I') = \ell((I', x_2', \ldots, x_n')/(x_2', \ldots, x_n')) = \ell((I, x_1, \ldots, x_n)/(x_1, \ldots, x_n))$. Since $\ell(I) = \ell(I')$, we are done.

PROPOSITION 6.22. Let I be an ideal in a local ring (R,M), and let $x_1, \ldots, x_n$ be an asymptotic sequence over I. Then $\ell(I) = \ell((I, x_1, \ldots, x_n)/(x_1, \ldots, x_n))$.

Proof: Suppose $n = 1$. Assume momentarily that R is a complete local domain, hence analytically unramified. Thus for large m, $J = \overline{I^m}$ is normal (that is, all powers are integrally closed) [SO]. We easily see that $\ell(I) = \ell(J)$ and $\overline{A}^*(I) = \overline{A}^*(J) = A^*(J)$. Since $x_1 \notin \cup\{P \in \overline{A}^*(I)\}$, by Lemma 6.21 we have $\ell(J) = \ell((J, x_1)/(x_1))$. Now $J = \overline{I^m}$ implies $((I, x_1)/(x_1))^m \subseteq (J, x_1)/(x_1) \subseteq \overline{((I, x_1)/(x_1))^m}$, from which we see that $\ell((I, x_1)/(x_1)) = \ell((J, x_1)/(x_1)) = \ell(J) = \ell(I)$, completing this case. Now assume $n = 1$ and (R,M) is arbitrary. Since $\ell(I) = \ell(IR^*)$ and $\ell((I, x_1)/(x_1)) = \ell((I, x_1)R^*/x_1R^*)$, we may suppose that R is complete. By Lemma 4.2, for some minimal prime q, $\ell(I) = \ell(I+q/q)$. In the complete local domain R/q, we get $\ell(I) = \ell(I+q/q) = \ell((I, q, x_1)/(q, x_1)) \leq \ell((I, x_1)/(x_1)) \leq \ell(I)$, by two uses of Lemma 6.18. This finishes the case $n = 1$.

Now inductively assume the result holds for $x_1, \ldots, x_{n-1}$. We first treat the case that (R,M) is quasi-unmixed. In this case, Lemma 6.19 and Proposition 6.20

show that x_n' is an asymptotic sequence over I', the primes denoting modulo $(x_1, \ldots, x_{n-1})$. By the case $n = 1$, $\ell(I') = \ell((I', x_n')/(x_n)') = \ell((I, x_1, \ldots, x_n)/(x_1, \ldots, x_n))$. Since our induction shows that $\ell(I) = \ell(I')$, we have proved our result for R quasi-unmixed.

In general, we have $\ell(I) = \ell(IR^* + q/q)$ for an appropriate minimal prime q of R^*. Since $x_1 + q, \ldots, x_n + q$ is an asymptotic sequence over $IR^* + q/q$, and since R^*/q is quasi-unmixed, the preceding shows that $\ell(I) = \ell((I, x_1, \ldots, x_n, q)R^*/(x_1, \ldots, x_n, q)R^*) \leq \ell((I, x_1, \ldots, x_n)R^*/(x_1, \ldots, x_n)R^*) = \ell((I, x_1, \ldots, x_n)/(x_1, \ldots, x_n)) \leq \ell(I)$, using Lemma 6.18. This completes the proof.

COROLLARY 6.23. Let I be an ideal in a locally quasi-unmixed Noetherian ring. Let $y_1, \ldots, y_m$ be an asymptotic sequence over I, and let primes denote modulo $(y_1, \ldots, y_m)$. Then $P' \in \overline{A}^*(I')$ if and only if $P \in \overline{A}^*(I, y_1, \ldots, y_m)$.

Proof: We may localize at P. Since Proposition 6.20 tells us that $y_1, \ldots, y_m$ is an asymptotic sequence, we use Lemma 6.19 for one direction. Thus suppose that $P \in \overline{A}^*(I, y_1, \ldots, y_m)$. By Proposition 4.1, height $P = \ell(I, y_1, \ldots, y_m) = \ell(I) + m$, using Corollary 6.2. Now Lemma 5.2 shows that height P' = height $P - m = \ell(I) = \ell(I')$, the last equality by Proposition 6.22. Thus Proposition 4.1 gives $P' \in \overline{A}^*(I')$.

When studying $\overline{A}^*(I)$ it obviously does no harm to replace I by any power of I, or by any ideal having the same integral closure as I. Therefore if as defined below, $I \sim J$, then $\overline{A}^*(I) = \overline{A}^*(J)$.

DEFINITION. Let I and J be ideals in a Noetherian ring. I and J are projectively equivalent $(I \sim J)$ means for some positive integers n and m, $\overline{I^n} = \overline{J^m}$.

Projective equivalence was studied by Samuel in [Sm] and in Chapter 10 we will have an interesting use for projective equivalence. For now, we present the following.

PROPOSITION 6.24. Let I and J be projectively equivalent. If $x_1, \ldots, x_n$ is an asymptotic sequence over I, then it is an asymptotic sequence over J.

Proof: As $\overline{A}^*(I) = \overline{A}^*(J)$, the statement is true for x_1. Assume it true for $x_1, \ldots, x_{n-1}$. To show it true for $x_1, \ldots, x_n$, it suffices to show that $\overline{A}^*(I, x_1, \ldots, x_{n-1}) = \overline{A}^*(J, x_1, \ldots, x_{n-1})$. Suppose $P \in \overline{A}^*(I, x_1, \ldots, x_{n-1})$. We may assume that R is local at P, and for an appropriately chosen minimal prime q of R^*, we have $P^*/q \in \overline{A}^*((I, x_1, \ldots, x_{n-1})R^* + q/q)$. As R^*/q is quasi-unmixed, height $P^*/q = \ell((I, x_1, \ldots, x_{n-1})R^* + q/q) = \ell(IR^* + q/q) + n - 1$, using Corollary 6.2. Now since $\overline{I^m} = \overline{J^k}$ for some positive integers m and k, it is straightforward to see that $\ell(IR^* + q/q) = \ell(JR^* + q/q)$. Thus height $P^*/q = \ell(JR^* + q/q) + n - 1 = \ell((J, x_1, \ldots, x_{n-1})R^* + q/q)$, since our inductive assumption shows that $x_1 + q, \ldots, x_{n-1} + q$ is an asymptotic sequence over $JR^* + q/q$. Therefore $P^*/q \in \overline{A}^*((J, x_1, \ldots, x_{n-1})R^* + q/q)$ and so $P \in \overline{A}^*(J, x_1, \ldots, x_{n-1})$. The reverse is identical.

CHAPTER VII: Asymptotic Grade

In [B2] Brodmann shows that if I is an ideal in a local ring (R,M), then the sequence $\text{grade}(R/I^n)$ $n = 1,2,\ldots$ eventually becomes constant, and if this constant is β, then $\ell(I) \leq \text{height } M - \beta$. This chapter presents these proofs and tries (with only limited success) to extend them to $\text{grade}(R/\overline{I^n})$. (Earlier work of Burch [Bu] has shows that $\ell(I) \leq \text{height } M - \liminf \text{grade}(R/\overline{I^n})$, at least when R/M is infinite.)

Our first proposition will be an extension of Proposition 1.3 to modules, and as the proof is essentially identical, it will be omitted.

PROPOSITION 7.1. Let $T = \Sigma R_n$, $n \geq 0$, be a Noetherian homogeneous graded ring, and let $A = \Sigma A_n$, $n \geq 0$, be a finitely generated graded T-module. Then for large n, $\text{Ass}_{R_0}(A_n) = \text{Ass}_{R_0}(A_{n+1}) = \ldots$.

The next result is due to Brodmann, generalized somewhat by D. Katz in order to glean some information about $\text{grade}(R/\overline{I^n})$.

PROPOSITION 7.2. Let R be a Noetherian ring, and let A be a finitely generated R module. Let $A = A_0 \supseteq A_1 \supseteq A_2 \supseteq \ldots$ be a sequence of submodules, and suppose that there is an ideal I of R and an integer $h \geq 0$ such that for all $n \geq 0$ $I^{n-h}A \supseteq A_n \supseteq A_{n+1} \supseteq IA_n$. (Here, $I^{n-h} = R$ when $n \leq h$.) Then

(i) $\text{Ass}_R(A_{n-1}/A_n)$ stabilizes for large n.

(ii) $\text{Ass}_R(A/A_n)$ stabilizes for large n.

(iii) If R is local, then $\text{grade}_R(A/A_n)$ stabilizes for large n, and this stable value is equal to or less than $\text{grade}_R A$.

(iv) If R is a local ring and if n is large enough that $\text{grade}_R(A/A_n)$ has stabilized, say at β, then if $x_1, \ldots, x_\beta$ is an R-sequence on A/A_n, it is also an R-sequence on A/A_k for all $k \geq n$.

Proof: With X an indeterminate, let $T = R + IX + I^2X^2 + \ldots$, and let $B = A_0 + A_1X + A_2X_2 + \ldots$, (in both cases meaning finite polynomials). B is a T-module. The hypothesis shows that $A_h + A_{h+1}X + A_{h+2}X^2 + \ldots$ is contained in $A + IAX + I^2AX^2 + \ldots$. This last is a graded T-module, finitely generated by the same elements which generate A over R. Since $T=R[IX]$ is Noetherian, $A_h + A_{h+1}X + A_{h+2}X^2 + \dots$ is a finitely generated T-module. We now easily see that B is a finitely generated T-module. Since $A_1 + A_2X + A_3X^2 + \dots$ is a submodule of B, we have that $A/A_1 + A_1/A_2\, X + A_2/A_3\, X^3 + \dots$ is a finitely generated T-module. Using Proposition 7.1, since $T_0 = R$, we have (i).

For (ii), consider the exact sequence $0 \to A_{n-1}/A_n \to A/A_n \to A/A_{n-1} \to 0$, which shows that $\mathrm{Ass}_R(A/A_n) \subseteq \mathrm{Ass}_R(A/A_{n-1}) \cup \mathrm{Ass}_R(A_{n-1}/A_n)$. Since for large n, $\mathrm{Ass}_R(A_{n-1}/A_n) = \mathrm{Ass}_R(A_{n-2}/A_{n-1}) \subseteq \mathrm{Ass}_R(A/A_{n-1})$, we have $\mathrm{Ass}_R(A/A_n) \subseteq \mathrm{Ass}_R(A/A_{n-1})$. As these sets are finite, they must stabilize, proving (ii).

For (iii), let (R,M) be a local ring, and suppose that m is large enough that $\mathrm{Ass}_R(A/A_m)$ has stabilized. First note that if $x \in M - \cup\{Q \in \mathrm{Ass}(A/A_m)\}$ then x is not a zero divisor on A. To see this, suppose that $xa = 0$. The choice of x and m shows that $a \in A_k$ for all $k \geq m$. However, the hypothesis shows that $\cap A_k \subseteq \cap I^{k-h}A = 0$ for $k = 1,2,\dots,$ so that $a = 0$ as desired.

We now induct on $\mathrm{grade}_R A$. If $\mathrm{grade}_R A = 0$, the preceding paragraph shows that $M \in \mathrm{Ass}(A/A_k)$ for all $k \geq m$, so that $\mathrm{grade}_R(A/A_k)$ stabilizes at 0. Now suppose that $\mathrm{grade}_R A > 0$. If $M \in \mathrm{Ass}_R(A/A_m)$, then again $\mathrm{grade}(A/A_k)$ stabilizes at 0. Now say $x \in M - \cup\{Q \in \mathrm{Ass}_R(A/A_m)\}$. By the preceding paragraph, x is not a zero divisor on A, and so letting $A_k' = A_k + xA/xA$, we have $\mathrm{grade}_R A' = \mathrm{grade}_R A - 1$. Also the sequence $A' = A_0' \supseteq A_1' \supseteq \dots$ satisfies the hypothesis of the theorem, and so by induction, $\mathrm{grade}_R(A'/A_k')$ stabilizes to a value not exceeding $\mathrm{grade}_R A' = \mathrm{grade}_R A - 1$. However, $A'/A_k' \approx A/A_k + xA \approx (A/A_k)/x(A/A_k)$. Since x is not a zero divisor on A/A_k, $k \geq m$, $\mathrm{grade}_R(A'/A_k') = \mathrm{grade}_R(A/A_k) - 1$. Thus $\mathrm{grade}_R(A/A_k)$ stabilizes to a value not exceeding $\mathrm{grade}_R A$.

For (iv), we induct on β, the case $\beta = 0$ being trivial. Now suppose that $\mathrm{grade}_R(A/A_k) = \beta > 0$ for all $k \geq n$, and let $x_1, \dots, x_\beta$ be an R-sequence on A/A_n. Letting $A_k' = A_k + x_1A/x_1A$, we have $A'/A_k' \approx (A/A_k)/x_1(A/A_k)$ so that for $k \geq n$, $\mathrm{grade}_R(A'/A_k') = \mathrm{grade}_R(A/A_k) - 1 = \beta - 1$. Thus $\mathrm{grade}_R(A'/A_k')$ has stabilized for $k \geq n$, and since our isomorphism shows that $x_2, \dots, x_\beta$ is an R-sequence on A'/A_n', by induction it is also an R-sequence on A'/A_k' for $k \geq n$. Again using our isomorphism, we get (iv).

Remark: The proof of (iii) above is a bit misleading. It does not show that $\mathrm{grade}_R(A/A_n)$ stabilizes as soon as $\mathrm{Ass}_R(A/A_n)$ stabilizes.

Question: Concerning (iv) above, is there some n such that if $x_1, \ldots, x_\beta$ is an R-sequence on A/A_k for some $k \geq n$, then it is also an R-sequence on A/A_n?

COROLLARY 7.3. (Brodmann [B2]) Let I be an ideal in a local ring (R,M). Then $\mathrm{grade}(R/I^n)$ stabilizes to a value not exceeding grade R.

Proof: Immediate, letting $A_n = I^n$, $h = 0$.

As mentioned previously, we do not know if this corollary holds with I^n replaced by $\overline{I^n}$. However, in two instances it does.

COROLLARY 7.4. (Katz) Let (R,M) be an analytically unramified local ring, and let I be an ideal. Then $\mathrm{grade}(R/\overline{I^n})$ stabilizes for large n.

Proof: Let $A = R$ and for $n \geq 1$, $A_n = \overline{I^n}$. As R is analytically unramified, there is an h with $I^{n-h} \supseteq \overline{I^n}$ for all $n > h$. Use Proposition 7.2.

COROLLARY 7.5. (Katz) Let (R,M) be a complete local ring, and let I be an ideal. Then $\mathrm{grade}(R/\overline{I^n})$ stabilizes.

Proof: Let $Z = \mathrm{rad}\, R$. Thus R/Z is analytically unramified. Letting $A = R/Z$ and for $n \geq 1$, $A_n = \overline{I^n}/Z$ (noting that $Z \subseteq \overline{I^n}$), and considering the ideal $I+Z/Z$ of R/Z, there is an $h \geq 0$ with $(I+Z/Z)^{n-h} \supseteq \overline{(I+Z/Z)^n} \supseteq A_n \supseteq A_{n+1} \supseteq (I+Z/Z)A_n$. By Proposition 7.2, $\mathrm{grade}_{R/Z}(A/A_n)$ stabilizes. Thus $\mathrm{grade}_R(R/\overline{I^n})$ stabilizes.

Question: If I is an ideal in a local ring (R,M), will $\mathrm{grade}(R/\overline{I^n})$ always stabilize?

By Proposition 7.2 (iv), we see that it makes sense to discuss elements $x_1, \ldots, x_s$ of R whose images in R/I^n form an R-sequence for all large n. We now show that such elements form an R-sequence, an asymptotic sequence, and if R is locally quasi-unmixed, an asymptotic sequence over I.

PROPOSITION 7.6. Let (R,M) be a local ring, and let I be an ideal. Let $x_1,\ldots,x_s$ be elements of R whose images in R/I^n are an R-sequence for all large n. Then $x_1,\ldots,x_s$ is an R-sequence.

Proof: We induct. Suppose $x_1y=0$. Then $x_1y \in I^n$ for all large n, and by the nature of x_1, $y \in \cap I^n = 0$. Thus x_1 is an R-sequence. Letting primes denote modulo x_1, since $R'/I'^n \approx R/(I^n,x_1) \approx (R/I^n)/x_1(R/I^n)$, we see that the images of $x_2',\ldots,x_s'$ in R'/I'^n is an R-sequence. By induction, $x_2',\ldots,x_s'$ in R', so that $x_1,\ldots,x_s$ is an R-sequence in R.

Note that since the $x_1,\ldots,x_s$ of Lemma 7.6 are an R-sequence, Lemma 5.13 shows they are an asymptotic sequence.

PROPOSITION 7.7. Let I be an ideal in a locally quasi-unmixed Noetherian ring, and suppose that the images of $x_1,\ldots,x_s$ are an R-sequence in R/I^n for all large n. Then $x_1,\ldots,x_s$ are an asymptotic sequence over I.

Proof: Since x_1 is not a zero divisor on R/I^n for all large n, we see that $x_1 \notin \cup\{Q \in A^*(I)\}$. As $\bar{A}^*(I) \subseteq A^*(I)$, we have x_1 an asymptotic sequence over I. We now inductively assume that $x_1,\ldots,x_{s-1}$ is an asymptotic sequence over I, and we must show that $x_s \notin \cup\{P \in \bar{A}^*(I,x_1,\ldots,x_{s-1})\}$. Suppose to the contrary that $x_s \in P \in \bar{A}^*(I,x_1,\ldots,x_{s-1})$. We may localize at P, and invoke Proposition 4.1 and Corollary 6.2 to see that height $P = \ell(I)+s-1$. Letting a prime denote modulo $(x_1,\ldots,x_{s-1})$, Lemma 6.21 gives $\ell(I') = \ell(I) =$ height $P-(s-1)$. Since (R,P) is quasi-unmixed, and since Proposition 7.6 shows that height$(x_1,\ldots,x_{s-1}) = s-1$, we have height $P' =$ height $P-(s-1) = \ell(I')$. By Proposition 4.1, $P' \in \bar{A}^*(I')$. (We could have simply invoked Corollary 6.23, but that uses Proposition 6.22, which really is not needed.) Since $\bar{A}^*(I') \subseteq A^*(I')$, for large n we have $P' \in \text{Ass}(R'/I'^n)$, so that $P \in \text{Ass}(R/(I^n,x_1,\ldots,x_{s-1}))$. Since $x_s \in P$, we have contradicted that $x_1,\ldots,x_s$ is an R-sequence modulo I^n for large n.

Question: If the images of $x_1,\ldots,x_s$ are an R-sequence in R/I^n for all large n, is $x_1,\ldots,x_s$ an asymptotic sequence over I even when R is not locally quasi-unmixed?

Question: Let (R,M) be a local ring and let I be an ideal. Let r be the length of a maximal asymptotic sequence over I, and let β be the eventual stable value of $\text{grade}(R/I^n)$. Is $\beta \leq r$. (Yes, if R is quasi-unmixed, by Proposition 7.7.)

Remark: If the answer to this last question is yes, then our next result would follow from Corollary 6.4. The proof we give is due to Brodmann.

PROPOSITION 7.8. Let I be an ideal in a local ring (R,M). Let β be the eventual stable value of $\text{grade}(R/I^n)$. Then $\beta \leq \text{height } M - \ell(I)$.

Proof: We induct on β, the case $\beta = 0$ following from Lemma 4.3. For $\beta > 0$, we have $M \notin A^*(I)$. Pick $x \in M - \cup \{Q \in A^*(I)\}$, and let a prime denote modulo x. By Lemma 6.21 we have $\ell(I) = \ell(I')$. Also, Lemma 7.6 shows that x is not a zero divisor in R, so that $\text{height } M' = \text{height } M - 1$. Finally, $R'/I'^n \approx R/(I^n,x) \approx (R/I^n)/x(R/I^n)$ so that $\beta' = \text{grade } R'/I'^n = \text{grade } R/I^n - 1 = \beta - 1$. By induction, $\beta' \leq \text{height } M' - \ell(I')$ and therefore $\beta \leq \text{height } M - \ell(I)$.

In Proposition 7.8, equality need not occur. Let (R,M) be a 2-dimensional local domain which is not Cohen-Macaulay and let I be principal. Then $\ell(I) = 1$ and $\beta = 0$.

Let (R,M) be a local domain with R/M infinite, and let I be an ideal. In [Bu], Burch shows that $\ell(I) \leq \text{height } M - \text{lin inf grade}(R/I^n)$. Propositions 7.2 and 7.8 give a shorter proof, and also show that the lin inf is in fact the limit β. However, [Bu] also shows that $\ell(I) \leq \text{height } M - \text{lin inf grade}(R/\overline{I^n})$. As stated earlier, we do not know if $\text{grade}(R/\overline{I^n})$ always stabilizes. When it does, (such as in Corollaries 7.4 and 7.5), we can combine that fact with Burch's second inequality to get the following.

PROPOSITION 7.9. Let I be an ideal in a local ring (R,M) with R/M infinite, and suppose that $\text{grade}(R/\overline{I^n}) = \overline{\beta}$ for all large n. Then $\overline{\beta} \leq \text{height } M - \ell(I)$.

In Proposition 7.9, equality need not occur. Let (R,M) be a 3-dimensional normal Noetherian domain which is not Cohen-Macaulay. Let $I = aR \neq 0$ be principal. As R is normal, $\overline{I^n} = \overline{(a^n)} = (a^n)$. Thus $\text{grade}(R/\overline{I^n}) \neq 2$ or else we would have grade $R = 3$ making R Cohen-Macaulay. If for any n, $\mathrm{grade}\,(R/\overline{I^n}) = 0$, then $M \in \overline{A}(I,n) \subseteq \overrightarrow{A}^*(I)$ showing that $\mathrm{grade}\,(R/\overline{I^k}) = 0$ for all large k, and $\overline{\beta}$ exists and equals 0. Otherwise, we have $\mathrm{grade}\,(R/\overline{I^n})$ is neither 2 nor 0, and so in this case $\overline{\beta}$ exists and is 1. Thus $\overline{\beta}$ is either 0 or 1, while $\ell(I) = 1$ and height $M = 3$.

Question: With β as in Proposition 7.8 and $\overline{\beta}$ as in Proposition 7.9, is $\overline{\beta} \geq \beta$? (Note that $\overline{\beta} = 0$ implies $M \in \overline{A}^*(I) \subseteq A^*(I)$ giving $\beta = 0$.)

While we do not know if the $\overline{\beta}$ of Proposition 7.9 always exists, we may go to the completion, and use Corollary 7.5. Since height $M =$ height M^* and $\ell(I) = \ell(IR^*)$ we have the following.

PROPOSITION 7.10. Let (R,M) be a local ring and let I be an ideal. Let $\overrightarrow{\beta}^* = \mathrm{grade}\,(R^*/\overline{I^nR^*})$ for all large n. Then $\overrightarrow{\beta}^* \leq$ height $M - \ell(I)$.

Question: Note that $\mathrm{grade}\,R/\overline{I^n} = \mathrm{grade}\,R^*/\overline{I^nR^*}$. How does $\mathrm{grade}\,R/\overline{I^n}$ compare to $R^*/\overline{I^nR^*}$?

CHAPTER VIII: When $A^* = \vec{A}^*$

In Corollary 4.7, we saw that in a 2-dimensional normal Noetherian domain, $A^*(I) = \vec{A}^*(I)$ for any ideal I. In this chapter, we will attempt to identify all domains having that property, and will almost succeed. In particular, among domains satisfying the altitude formula, we will see that Corollary 4.7 tells the whole story.

The path we will take generally follows that laid out by Ratliff in [R6]. However, we will take some shortcuts by using Lemma 8.3.

Notation: Let $\mathcal{A}$ denote the class of Noetherian domains in which $A^*(I) = \vec{A}^*(I)$ for every ideal I.

We note that the Noetherian domain R is in $\mathcal{A}$ if and only if $R_P \in \mathcal{A}$ for every prime P of R. Therefore we will content ourselves with investigating local domains in $\mathcal{A}$.

We also note that if $J \subseteq I \subseteq \bar{J}$ are ideals, then for all n, J^n reduces I^n, so that $J^n \subseteq I^n \subseteq \overline{J^n}$. Thus $\vec{A}^*(I) = \vec{A}^*(J)$.

LEMMA 8.1. Let I be an ideal in a Noetherian ring R. Let $x \in I$ be a regular element. Then for large n, $(I^{n+h} : I^h) = I^n$, for all h.

Proof: Consider t as in Lemma 1.1(b). Now $x(I^{n+1} : x) = I^{n+1} \cap (x)$, and by the Artin-Rees lemma, there is an integer $r > 0$ such that for large n, $I^{n+1} \cap (x) = I^{n+1-r}(I^r \cap (x)) \subseteq xI^{n+1-r}$. Thus $(I^{n+1} : I) \subseteq (I^{n+1} : x) \subseteq I^{n+1-r} \subseteq I^t$, since x is regular. By Lemma 1.1(b), for large n we now have $(I^{n+1} : I) = I^n$, and so we easily see that $(I^{n+h} : I^h) = I^n$ for all h.

Notation: Let I be a regular ideal in a Noetherian ring. Since $(I^2 : I) \subseteq (I^3 : I^2) \subseteq (I^4 : I^3) \subseteq \ldots$ eventually stabilizes, we let $\tilde{I} = (I^{n+1} : I^n)$ for all large n.

$\tilde{I}$ was introduced by Ratliff and Rush [RR].

LEMMA 8.2. Let I be a regular ideal in a Noetherian ring.

(i) $\tilde{I}^m = I^m$ for all large m.

(ii) $\widetilde{I^m} = I^m$ for all large m.

(iii) $\tilde{\tilde{I}} = \tilde{I}$.

(iv) Let J be an ideal containing I. The following are equivalent.

(a) $J \subseteq \tilde{I}$.

(b) $\tilde{J} = \tilde{I}$.

(c) For all large m, $I^m = J^m$.

(d) For some $m \geq 1$, $I^m = J^m$.

(v) Let J be any ideal and suppose that for some m, $J^m = I^m$ and $J^{m+1} = I^{m+1}$. Then $J \subseteq \tilde{I}$.

(vi) $I \subseteq \tilde{I} \subseteq \bar{I}$.

Proof: (i): For large m, $\tilde{I} = (I^{m+1} : I^m)$, so $\tilde{I}^m \subseteq (I^{m^2+m} : I^{m^2}) = I^m$, by Lemma 8.1. As $I \subseteq \tilde{I}$, (i) is clear.

(ii): Using Lemma 8.1, for large m and any h, $(I^{m(h+1)} : I^{mh}) = I^m$. Thus $\widetilde{I^m} = I^m$.

(iii): For large m, $\tilde{\tilde{I}} = (\tilde{I}^{m+1} : \tilde{I}^m)$. Using (i), this equals $(I^{m+1} : I^m) = \tilde{I}$.

(iv): (a) $\Rightarrow$ (c): Since $I \subseteq J \subseteq \tilde{I}$, (i) shows that $I^m = J^m$ for all large m. (c) $\Rightarrow$ (b): For large m, $\tilde{I} = (I^{m+1} : I^m) = (J^{m+1} : J^m) = \tilde{J}$. (b) $\Rightarrow$ (d): Combine (b) with (i). (d) $\Rightarrow$ (a): Suppose $I^m = J^m$. Then $JI^{m-1} \subseteq J^m = I^m$, so $J \subseteq (I^m : I^{m-1}) \subseteq \tilde{I}$.

(v) $JI^m = JJ^m = J^{m+1} = I^{m+1}$. Thus $J \subseteq (I^{m+1} : I^m) \subseteq \tilde{I}$. (Note: Unlike in (iv)d, $I^m = J^m$ is not enough. Let $R = K[X,Y]$ modulo $(X^2 - Y^4)$. Let $I = (X)$ and $J = (Y^2)$. Then $J^2 = I^2$ but $J \not\subseteq \tilde{I} = I$.)

(vi): Obviously $I \subseteq \tilde{I}$. Now if $x \in \tilde{I}$, then (i) shows $x^m \in I^m$ for large m. Hence $x \in \bar{I}$.

LEMMA 8.3. Let (R,M) be a local ring. Let $I \subseteq J$ be ideals with $J \not\subseteq \tilde{I}$. Then $M \in \mathrm{Ass}(R/(I+MJ)^n)$ for all $n \geq 1$. In particular, $M \in A^*(I+MJ)$.

Proof: By Lemma 8.2(iv), for all $n \geq 1$ we have $I^n \neq J^n$. Thus by Nakayama's lemma, $J^n \not\subseteq I^n + MJ^n$. Since clearly $(I+MJ)^n \subseteq I^n + MJ^n$, we have $J^n \not\subseteq (I+MJ)^n$. However, $M^n J^n \subseteq (I+MJ)^n$, showing that M^n consists of zero divisors modulo $(I+MJ)^n$. As M is maximal, $M \in \mathrm{Ass}(R/(I+MJ)^n)$.

LEMMA 8.4. Let (R,M) be a local domain which is not normal. Then for some $0 \neq a \in R$, $M \in A^*((a) + M(\overline{a}))$.

Proof: Pick $\alpha \in \overline{R} - R$ and $0 \neq a \in R$ with $a\alpha \in R$. Then $a\alpha \in a\overline{R} \cap R = \overline{aR}$ but $a\alpha \notin aR$. Thus $(a) \neq (\overline{a})$. However we clearly have $(a) = (\tilde{a})$. Therefore $(\overline{a}) \not\subseteq (\tilde{a})$ and we use Lemma 8.3.

LEMMA 8.5. Let a,b be a pair of analytically independent elements in the local domain (R,M). Then $M \in A^*((a^2,b^2) + M\overline{(a^2,b^2)})$.

Proof: Let $I = (a,b)$ and $J = (a^2,b^2) \subseteq I^2$. Since a,b are analytically independent, the minimal number of generators of I^{2n}, for $n \geq 1$, is $2n+1$. As a^2, b^2 are also analytically independent, the minimal number of generators of J^n is $n+1$. Hence $J^n \neq I^{2n}$ for any n, and so by Lemma 8.2(iv) $I^2 \not\subseteq \tilde{J}$. However, since ab satisfies $X^2 - a^2b^2$, we have $J \subseteq I^2 \subseteq \overline{J}$. Thus $\overline{J} \not\subseteq \tilde{J}$, and so we use Lemma 8.3.

LEMMA 8.6. Let R be a Noetherian domain with $R \in \mathcal{Q}$. Let a,b be elements of R such that no height 1 prime of R contains (a,b). Then if $J = (a^2,b^2)$, $\overrightarrow{A}^*(J) = \{P \text{ prime} \mid (a,b) \subseteq P\}$. In particular, only finitely many primes of R contain (a,b).

Proof: One inclusion is obvious. Thus suppose $(a,b) \subseteq P$. As no height 1 prime of R_P contains (a,b), a and b are analytically independent in R_P. By Lemma 8.5, $P_P \in A^*(J_P + P_P\overline{J}_P)$. Thus $P \in A^*(J + P\overline{J}) = \overrightarrow{A}^*(J + P\overline{J}) = \overrightarrow{A}^*(J)$, using that $R \in \mathcal{Q}$ and that $J \subseteq J + P\overline{J} \subseteq \overline{J}$.

PROPOSITION 8.7. If $R \in \mathcal{Q}$, then $\dim R \leq 3$.

Proof: If $\dim R > 3$, there is a height 2 prime P with $\operatorname{depth} P > 1$. Pick a,b in P with no height 1 prime containing both a and b. As $\operatorname{depth} P > 1$, infinitely many primes contain P, and hence contain (a,b). This contradicts Lemma 8.6.

We now proceed in two cases, first assuming that (R,M) is quasi-unmixed.

PROPOSITION 8.8. Let the local domain (R,M) be quasi-unmixed. Then $R \in \mathcal{C}$ if and only if either R is 1-dimensional or R is 2-dimensional and normal.

Proof: If $\dim R = 1$, then for any $I \neq 0$, $A^*(I) = \{M\} = \overline{A}^*(I)$. If $\dim R = 2$ and R is normal, use Corollary 4.7.

Conversely, suppose that $R \in \mathcal{C}$ and $\dim R > 1$. We will show that R is 2-dimensional and normal. Since $\dim R > 1$, choose a,b analytically independent in R. By Lemma 8.5, $M \in A^*((a^2,b^2) + M\overline{(a^2,b^2)})$. Since $R \in \mathcal{C}$ and since $(a^2,b^2) \subseteq (a^2,b^2) + M\overline{(a^2,b^2)} \subseteq \overline{(a^2,b^2)}$, we see that $M \in \overline{A}^*(a^2,b^2)$. By Proposition 4.1, height $M = \ell(a^2,b^2) = 2$. Thus $\dim R = 2$. Now suppose that R is not normal. By Lemma 8.4, for some $a \in R$ we have $M \in A^*((a) + M(\overline{a})) = \overline{A}^*((a) + M(\overline{a})) = \overline{A}^*((a))$. By Proposition 4.1, height $M = \ell((a)) = 1$, a contradiction. Thus R is normal.

PROPOSITION 8.9. Let (R,M) be a local domain which is not quasi-unmixed (so $\dim R > 1$).

(i) If $\dim R = 2$, $R \in \mathcal{C}$

(ii) If $\dim R = 3$ and if $R \neq \overline{R}$, then $R \in \mathcal{C}$ if and only if $\overline{R}$ contains a height 1 maximal, and for all height 2 primes P of R, either R_P is normal or $\overline{R}$ contains a height 1 prime lying over P.

Proof: (i) Since the only way a 2-dimensional local domain can fail to be quasi-unmixed is for $\overline{R}$ to contain a height 1 maximal, by Proposition 3.19 we see that $M \in \overline{A}^*(I)$ for all $I \neq 0$. As $\overline{A}^*(I) \subseteq A^*(I)$, we now easily see that $\overline{A}^*(I) = A^*(I)$.

(ii) Suppose that $R \in \mathcal{C}$ Since $R \neq \overline{R}$, Lemma 8.4 says $M \in A^*((a) + M(\overline{a})) = \overline{A}^*((a) + M(\overline{a})) = \overline{A}^*((a))$ for some $a \in R$. By Proposition 3.19, $\overline{R}$ contains a height 1 maximal. Now let P have height 2 in R. Of course $R_P \in \mathcal{C}$ If R_P is quasi-unmixed, Proposition 8.8 shows that it is normal. On the other hand, if R_P is not quasi-unmixed, then as mentioned in (i) above, $\bar{R}_P$ must contain a height 1 prime lying over P_P, so that $\bar{R}$ contains a height 1 prime lying over P.

For the converse, assume that R has the given properties. We will show that $R \in \mathcal{Q}$. Now Proposition 3.19 shows that $M \in \bar{A}^*(I) \subseteq A^*(I)$ for all $I \neq 0$. Also a height 1 prime is in $\bar{A}^*(I)$ if and only if it is minimal over I if and only if it is in $A^*(I)$. Therefore we need only consider a height 2 prime P and show $P \in \bar{A}^*(I)$ if and only if $P \in A^*(I)$. If R_P is normal, then it is Cohen-Macaulay, hence quasi-unmixed, and by Proposition 8.8, $P_P \in \bar{A}^*(I_P)$ if and only if $P_P \in A^*(I_P)$. If R_P is not normal, then by hypothesis, $\bar{R}$ contains a height 1 prime lying over P. By Proposition 3.19, $P \in \bar{A}^*(I) \subseteq A^*(I)$ for all $I \neq 0$. This completes the proof.

One case eludes our efforts. If (R,M) is a 3-dimensional normal local domain which is not quasi-unmixed, then we do not know if R must be, might be, or cannot be in $\mathcal{Q}$. (Only recently has it been shown that normal Noetherian domains which are not quasi-unmixed exist. See [O] or [H2].) The next proposition might help resolve the issue.

PROPOSITION 8.10. Let (R,M) be a 3-dimensional normal local domain which is not quasi-unmixed. Assume that R/M is infinite. Then $R \in \mathcal{Q}$ if and only if $M \in \bar{A}^*(a,b)$ for any pair of analytically independent elements a,b.

Proof: Note that $R_P \in \mathcal{Q}$ for any prime P of height 1 or 2, the first case being obvious and the second using Corollary 4.7. Thus for any I, the primes of height 1 or 2 in $A^*(I)$ are identical to those in $\bar{A}^*(I)$. Since $\bar{A}^*(I) \subseteq A^*(I)$, we see that the only way to have $A^*(I) \neq \bar{A}^*(I)$ is to have $M \in A^*(I) - \bar{A}^*(I)$. Now if $\ell(I) = 1$ then for some $a \in R$ we have $(a) \subseteq I \subseteq \overline{(a)}$. As R is normal, $(a) = \overline{(a)}$ so that $I = (a)$, and hence $M \notin A^*(I)$ since R is a Krull domain and height $M > 1$. That is, $A^*(I) = \bar{A}^*(I)$ whenever $\ell(I) = 1$. Now suppose $\ell(I) = 3$. By Proposition 4.1, $M \in \bar{A}^*(I)$. Thus $\bar{A}^*(I) = A^*(I)$ whenever $\ell(I) = 3$. Since $\ell(I) \leq$ height $M = 3$, the only remaining case is $\ell(I) = 2$. Let $J = (a,b)$ be a minimal reduction of I. Then a,b are analytically independent. Suppose now that for any analytically independent pair a,b, we had $M \in \overrightarrow{A}^*(a,b)$. Then $M \in \overrightarrow{A}^*(J) = \overrightarrow{A}^*(I)$ so that $\overrightarrow{A}^*(I) = A^*(I)$ whenever $\ell(I) = 2$, showing that $R \in \mathcal{a}$. Conversly, suppose $R \in \mathcal{a}$ and consider a,b analytically independent. By Lemma 8.5, $M \in A^*((a^2,b^2)+M\overline{(a^2,b^2)}) = \overrightarrow{A}^*((a^2,b^2)+M\overline{(a^2,b^2)}) = \overrightarrow{A}^*(a^2,b^2)$. However $(a^2,b^2) \subseteq (a,b)^2 \subseteq \overline{(a^2,b^2)}$ so that clearly $\overrightarrow{A}^*(a^2,b^2) = \overrightarrow{A}^*((a,b)^2) = \overrightarrow{A}^*(a,b)$. Thus $M \in \overrightarrow{A}^*(a,b)$ as desired.

Having investigated when $\overrightarrow{A}^*(I) = A^*(I)$ for all I, we now look at two results, the first assuming more, the second assuming less.

PROPOSITION 8.11. Let (R,M) be a 2-dimensional local domain. Then the following are equivalent.

(i) R is a U.F.D.

(ii) For any ideal I of R, $A(I,1) = A(I,2) = \ldots = A^*(I) = \overline{A}(I,1) = \overline{A}(I,2) = \ldots = \overrightarrow{A}^*(I)$.

(iii) For any ideal I, $\mathrm{Ass}(R/I) = \overrightarrow{A}^*(I)$.

Proof: (i) ⇒ (ii): Any prime minimal over I is in each set in (ii). Thus we need only worry about M. Claim: If J is an ideal, then M is a prime divisor of J if and only if J is not principal if and only if $\overline{J}$ is not principal. Applying this claim to $J = I^n$, $n = 1,2,\ldots$, shows that if M is in any set in (ii), then it is in all of them. To prove the claim, the first equivalence is an easy exercise using primary decomposition and the fact that R is a U.F.D. For the second equivalence, we have (using that R is normal) J not principal if and only if $\ell(J) > 1$ if and only if $\ell(\overline{J}) > 1$ if and only if $\overline{J}$ is not principal. This proves the claim, and so (i) ⇒ (ii).

(ii) ⇒ (iii) is immediate since A(I,1) is Ass(R/I).

(iii) ⇒ (i): We first claim that R is normal. If not, by Lemma 8.4, for some $0 \neq a \in R$ we have $M \in A^*((a)+M(\overline{a}))$. Thus, for large n, M is a prime divisor of $((a)+M(\overline{a}))^n$. By (iii) and the fact that $(a)^n \subseteq ((a)+M(\overline{a}))^n \subseteq (\overline{a})^n \subseteq \overline{(a^n)}$, we see that $M \in \overrightarrow{A}^*((a))$. Proposition 3.19 now shows that $M \in \overrightarrow{A}^*(I)$ for any ideal $I \neq 0$ in R. By (iii), $M \in \mathrm{Ass}(R/I)$ for any ideal $I \neq 0$ in R. This last is impossible since we may let $I = P$ be a height 1 prime. This contradiction shows that R is normal. Now if R is not a U.F.D., let Q be a height 1 prime of R which is not principal. By Corollary 4.7, $M \in \overline{A}^*(Q)$. By (iii), $M \in \mathrm{Ass}(R/Q)$. This is impossible. Thus R is a U.F.D.

PROPOSITION 8.12. Let R be a locally quasi-unmixed Noetherian ring. Then the following are equivalent

i) R is Cohen-Macaulay.

ii) $A^*(I) = \overline{A}^*(I)$ for every ideal of the principal class.

Proof: (i) $\Rightarrow$ (ii): Let I be in the principal class. The argument used in [K1, Theorem 125] shows that I can be generated by elements $x_1, \ldots, x_n$ with $\mathrm{height}(x_1, \ldots, x_i) = i$, $i = 1, \ldots, n$. As R is Cohen-Macaulay, $x_1, \ldots, x_n$ is clearly an R-sequence. Using [K1, Exercise 13, page 103] we easily see that $A^*(I) = \mathrm{Ass}(R/I) = \{P \mid P \text{ is minimal over } I\} \subseteq \overline{A}^*(I) \subseteq A^*(I)$. Thus (ii) holds.

(ii) $\Rightarrow$ (i): Suppose that R is not Cohen-Macaulay. Let P be a maximal prime with $\mathrm{height}\, P > \mathrm{gr}\, P = n$, and let $x_1, \ldots, x_n$ be an R-sequence with $P \in \mathrm{Ass}(R/I)$ with $I = (x_1, \ldots, x_n)$. Again using [K1, Exercise 13, page 103], we have $P \in A^*(I)$. However since $\ell(I_P)$ does not exceed the number of generators of I_P, we have $\mathrm{height}\, P > n \geq \ell(I_P)$. By Proposition 4.1, $P \notin \overline{A}^*(I)$. As $x_1, \ldots, x_n$ being an R-sequence shows that I is in the principal class, we are done.

CHAPTER IX: Conforming Relations

Recall that primes $P \subseteq Q$ in a domain R are said to satisfy going down if for any integral extension domain T and any prime q of T lying over Q, there is a prime p of T lying over P with $p \subseteq q$. Since going down is well known to hold for $P \subseteq Q$ if R is normal, one easily sees that in the above definition it suffices to consider the case $T = \overline{R}$, the integral closure of R.

PROPOSITION 9.1. Let P be prime in a Noetherian domain R. Let Q be a prime containing P and such that Q is minimal with respect to the property $P \subset Q$ fails going down. Then for any ideal I with $P \subseteq I \subseteq Q$, we have $Q \in \overrightarrow{A}^*(I)$.

Proof: We find a prime q of $\overline{R}$ lying over Q such that no prime contained in q lies over P. Shrink q to a prime q_1 minimal over $I\overline{R}$. Clearly no prime contained in q_1 can lie over P. Thus $P \subset q_1 \cap R$ fails going down. By the minimality of Q, $q_1 \cap R = Q$. Now use Proposition 3.5.

COROLLARY 9.2. Let P be a prime in a Noetherian domain R. If $\overrightarrow{A}^*(P) = \{P\}$, then $P \subseteq Q$ satisfies going down for any prime Q containing P.

Proof: Use Proposition 9.1 with $I = P$.

The converse of Corollary 9.2 fails. Let P be a non-principal height 1 prime in a 2-dimensional normal local domain (R,M). Normality implies that $P \subset M$ satisfies going down, while Corollary 4.7 shows that $M \in \overrightarrow{A}^*(P)$.

COROLLARY 9.3. Let $P \subset P'$ be primes in a Noetherian domain which fail going down. Then there is a prime $Q \in \overrightarrow{A}^*(P)$ with $P \subset Q \subseteq P'$ and $P \subset Q$ fails going down.

Proof: Shrink P' to a prime Q minimal with the property that $P \subset Q$ fails going down. By Proposition 9.1 we have $Q \in \overrightarrow{A}^*(P)$.

COROLLARY 9.4. Let P be a prime in a Noetherian domain R, and let W be a set of primes of R, each of which contains P, and such that $P \subset P'$ fails going down for all $P' \in W$. Then $\cap\{P' \in W\}$ strictly contains P.

Proof: For $P' \in W$, use Corollary 9.3 to find $Q \in \overrightarrow{A}^*(P)$ with $Q \subseteq P'$ and $P \subset Q$ failing going down. Obviously $P \neq Q$. Thus $\cap\{P' \in W\}$ contains $\cap\{Q \in \overrightarrow{A}^*(P) \mid P \neq Q\}$. As $\overrightarrow{A}^*(P)$ is finite, this last intersection strictly contains P.

Our next result sharpens our appreciation of Corollary 9.4 by applying it to a special situation.

DEFINITION. Let $P \subset Q$ be primes with height $Q/P = 1$. We say that Q is _directly above_ P.

COROLLARY 9.5. Let P be prime in a Noetherian domain. Then $P \subset Q$ satisfies going down for all but finitely many of the primes Q directly above P.

Proof: Since any infinite collection of primes directly above P intersect at exactly P, the result follows from Corollary 9.4.

We wish to give a result about going down whose proof has much in common with an older result about chain conditions. In order to exhibit the similarities between these two proofs, we will utilize some abstraction.

DEFINITION. Let Q be a prime ideal and let W be an infinite set of prime ideals each of which properly contains Q. If for any infinite subset W' of W, $\cap\{Q' \in W'\} = Q$ then we call (Q,W) a _conforming pair_.

LEMMA 9.6. Let I be an ideal in a Noetherian ring and let W^* be an infinite set of primes each of which contains I. Then there is a conforming pair (Q,W) with $I \subseteq Q$ and $W \subseteq W^*$.

Proof: Enlarge I to an ideal Q maximal with respect to being properly contained in infinitely many primes in W^*. It is straightforward to verify that Q is prime. If $W = \{Q' \in W^* \mid Q \subset Q'\}$, it is easy to see that (Q,W) is a conforming pair.

For primes $P \subseteq Q$, we will consider a relation which may or may not hold between P and Q, writing either $P * Q$ or $P \not* Q$, respectively.

DEFINITION. We will call $*$ a <u>conforming relation</u> if whenever $P \subseteq Q$ and (Q,W) is a conforming pair, then $P * Q$ if and only if $P * Q'$ for all but finitely many $Q' \in W$.

LEMMA 9.7. If $*$ is a conforming relation, then $\not*$ is a conforming relation.

Proof: Suppose that $P \not* Q$ and that (Q,W) is a conforming pair. Let $W' = \{Q' \in W \mid P * Q'\}$. We must show that W' is finite. If W' is infinite, then (Q,W') is clearly a conforming pair, and since $P * Q'$ for all $Q' \in W'$, we must have $P * Q$ since $*$ is a conforming relation. This contradiction shows that W' is finite. Similar reasoning shows that if $P \not* Q'$ for all but finitely many $Q' \in W$, then $P \not* Q$.

LEMMA 9.8. Let R be a Noetherian ring and let $*$ be a conforming relation. Let $P \subseteq I$ be ideals with P prime. Let $W = \{Q \in \text{spec } R \mid I \subseteq Q \text{ and } P * Q\}$. Then W has only finitely many minimal members (with respect to inclusion).

Proof: Let W' consist of all the minimal members of W. If W' is infinite, then Lemma 9.6 shows that there is a conforming pair (Q,W'') with $I \subseteq Q$ and $W'' \subseteq W'$. Since $P * Q'$ for all $Q' \in W''$, we must have $P * Q$. Thus $Q \in W$. This contradicts that Q is properly contained in the members of W'', since W'' consists of minimal members of W.

We now see that conforming relations are particularly well behaved.

PROPOSITION 9.9. Let P be prime in a Noetherian ring, and let $*$ be a conforming relation with $P * P$. Then there is a chain of ideals $P = I_0 \subset I_1 \subset \ldots \subset I_n$ with the following property: If $P \subseteq Q$ with Q prime, and if j is the largest subscript with $I_j \subseteq Q$, then $P * Q$ if and only if j is even.

Proof: We let $I_0 = P$ and inductively construct the chain. Suppose that I_m has been constructed, and assume m is even (the case that m is odd being symmetric). Let $W = \{Q' \in \text{spec } R \mid I_m \subseteq Q' \text{ and } P \not* Q'\}$. By Lemmas 9.7 and 9.8, W has only finitely many minimal members. We define I_{m+1} to be the intersection of those finitely many primes. Since I_m is (by induction) a finite intersection of primes q satisfying $P * q$, clearly I_m is proper in I_{m+1}. Thus our chain eventually stops. Suppose now that I_m is the largest ideal in our chain which is contained in Q, (and still assume m even). If $P \not* Q$, Then $Q \in W$ and so by construction we would have $I_{m+1} \subseteq Q$, a contradiction. Thus our chain has the stated property.

PROPOSITION 9.10. Let R be a Noetherian ring, and for each $P \in \text{spec } R$ suppose there is a chain of primes $P = I_0 \subset I_1 \subset \ldots \subset I_n$, $n = n(P)$. Define a relation in the following way: For $P \subseteq Q$ let j be the largest subscript with $I_j \subseteq Q$, and say $P * Q$ if and only if j is even. Then $*$ is a conforming relation.

Proof: Let $P \subseteq Q$ and let (Q,W) be a conforming pair. Let I_j be the largest ideal in the chain associated with P such that $I_j \subseteq Q$. By the definition of conforming pair, at most finitely many members of W can contain I_{j+1} (lest $I_{j+1} \subseteq Q$). Thus I_j is the largest ideal of our chain contained in Q' for all but finitely many $Q' \in W$. The result is now obvious.

Remark: Propositions 9.9 and 9.10 together characterize conforming relations with the property that $P * P$ for all $P \in \text{spec } R$. To accommodate $P \not* P$, we notice that if $*$ is a conforming relation and if $U \subseteq \text{spec } R$ and if $P \hat{*} Q$ is defined as $P * Q$ for $P \in U$ and $P \not* Q$ for $P \notin U$, then $P \hat{*} Q$ is a conforming relation. Thus we have characterized all conforming relations.

We now consider concrete examples.

DEFINITION. The primes $P \subseteq Q$ are said to satisfy <u>catanicity</u> if height $Q =$ height $P +$ height Q/P. (In [HM] this was called normality.)

LEMMA 9.11. Let (Q,W) be a conforming pair in a Noetherian ring. Then $Q \subset Q'$ satisfies catanicity for all but finitely many $Q' \in W$.

Proof: Let height $Q = n$ and let Q be minimal over $(x_1, \ldots, x_n)$. Let $Q = p_0, p_1, \ldots, p_m$ be all of the primes minimal over $(x_1, \ldots, x_n)$. Suppose that $W' = \{Q' \in W \mid Q \subset Q' \text{ fails catanicity}\}$. Then for $Q' \in W'$, since height $Q' >$

height Q'/Q + height Q = height $Q'/Q + n$, the principal ideal theorem shows that height $Q'/(x_1, \ldots, x_n)$ > height Q'/Q. Thus Q' must contain one of $p_1, \ldots, p_m$. This is true for all $Q' \in W'$. We have $p_1 \cap \ldots \cap p_m \subseteq \cap\{Q' \in W'\}$. Since $p_1 \cap \ldots \cap p_m \not\subseteq p_0 = Q$, the fact that (Q,W) is a conforming pair shows that W' must be finite.

PROPOSITION 9.12. Catanicity is a conforming relation in a Noetherian ring R.

Proof: Let $P \subseteq Q$ and let (Q,W) be a conforming pair. Letting a prime denote modulo P, if we let $W' = \{q' \mid q \in W\}$ then obviously (Q',W') is a conforming pair in R'. Let $\hat{W} = \{q \in W \mid$ either $Q \subset q$ fails catanicity or $Q' \subset q'$ fails catanicity$\}$. By Lemma 9.11 applied to both (Q,W) and (Q',W'), we see that $\hat{W}$ is finite.

Now let $q \in W - \hat{W}$. Then we have both height q = height q/Q + height Q and height q/P = height q/Q + height Q/P. Of course we also have height $Q \geq$ height Q/P + height P, equality holding exactly when $P \subseteq Q$ satisfies catanicity. Thus height q = height q/Q + height $Q \geq$ height q/Q + height Q/P + height P = height q/P + height P. Therefore, examining the one inequality, we see that $P \subseteq Q$ satisfies catanicity if and only if $P \subseteq q$ satisfies catanicity for any $q \in W - \hat{W}$.

To prove the result, we note that if $P \subseteq Q$ satisfies catanicity, then for all but finitely many $q \in W$, (the exceptions being in $\hat{W}$) $P \subseteq q$ satisfies catanicity. Conversely if $P \subseteq q$ satisfies catanicity for all but finitely many $q \in W$, then since $W - \hat{W}$ is infinite, $P \subseteq q$ satisfies catanicity for some $q \in W - \hat{W}$, and the preceding shows that $P \subseteq Q$ satisfies catanicity.

We now show that going down is a conforming relation. The next lemma, a strengthening of Corollary 9.5, is the analogue of Lemma 9.11.

LEMMA 9.13. Let (Q,W) be a conforming pair in a Noetherian domain. Then $Q \subset Q'$ satisfies going down for all but finitely many $Q' \in W$.

Proof: Let $W' = \{Q' \in W \mid Q \subset Q'$ fails going down$\}$. By Corollary 9.4, $\cap\{Q' \in W'\}$ strictly contains Q. By the definition of conforming pair, W' must be finite.

PROPOSITION 9.14. In a Noetherian domain, going down is a conforming relation.

Proof: Let $P \subseteq Q$ be primes and let (Q,W) be a conforming pair. Suppose that $P \subseteq Q$ satisfies going down. If $Q \subset Q'$ also satisfies going down, then obviously $P \subset Q'$ satisfies going down. Therefore Lemma 9.13 shows that $P \subseteq Q'$ satisfies going down for all but finitely many $Q' \in W$.

Now suppose that $P \subseteq Q'$ satisfies going down for all but finitely many $Q' \in W$. We may ignore the finitely many exceptions and assume $P \subseteq Q'$ satisfies going down for all $Q' \in W$. Our task is to show that $P \subseteq Q$ satisfies going down. Therefore, if q is a prime of $\bar{R}$ lying over Q, and if $p_1, \ldots, p_n$ are all the primes of $\bar{R}$ lying over P, we must show that q contains one of $p_1, \ldots, p_n$. By going up, for each $Q' \in W$, there is a q' prime in $\bar{R}$ with $q' \cap R = Q'$ and $q \subset q'$. Since $P \subseteq Q'$ satisfies going down, each such q' contains one of $p_1, \ldots, p_n$. Therefore some p_i is contained in infinitely many such q'. Let $U = \{q' \mid p_i \subseteq q'\}$. We claim that $\cap\{q' \in U\} = q$. This will show that $p_i \subseteq q$, completing the proof. To prove the claim, let $x \in \cap\{q' \in U\}$. Since x is integral over R, we may consider an expression $x^{\ell} + r_{\ell-1}x^{\ell-1} + \ldots + r_0 \in q$, with $r_i \in R$ and with ℓ minimal among such. Since this expression is in $q \subseteq q'$, and also $x \in q'$, we have $r_0 \in q' \cap R = Q' \in W$. As U is infinite, we now see that r_0 is contained in infinitely many members of W. Since (Q,W) is a conforming pair, $r_0 \in Q \subseteq q$. Thus $x(x^{\ell-1} + r_{\ell-1}x^{\ell-2} + \ldots + r_1) \in q$. As q is prime, the minimality of ℓ shows that $x \in q$ as desired.

Recall that for R Noetherian, the prime Q is a G-ideal exactly when either Q is maximal or $\dim(R/Q) = 1$ and R/Q has only finitely many prime ideals [K1, Theorem 146]. Recall also that the domain R satisfies going down if $P \subset Q$ satisfies going down for all primes $P \subset Q$.

PROPOSITION 9.15. Let R be a Noetherian domain and suppose that $P \subset Q$ satisfies going down whenever Q is a G-ideal and height $Q/P = 1$. Then R satisfies going down.

Proof: Suppose some $p \subset q$ does not satisfy going down. We may take a counter-example with q maximal among all such. We claim that q is a G-ideal. Let $W = \{q' \in \text{spec } R \mid q \subset q' \text{ and height}(q'/q) = 1\}$. If q is not a G-ideal, then W is infinite and (q,W) is a conforming pair. By the maximality of q, for any $q' \in W$, $p \subset q'$ has going down. By Proposition 9.14, $p \subset q$ must have going down, which is a contradiction. Thus q is a G-ideal.

Fixing q, we now may assume that p has been chosen to make height(q/p) minimal. If height$(q/p) > 1$ then consider the infinite set $U = \{p' \in \text{spec } R \mid p \subset p' \subset q \text{ and height}(p'/p) = 1\}$. By Corollary 9.5, there is a $p' \in U$ with $p \subset p'$ satisfying going down. Clearly $p' \subset q$ cannot satisfy going down. However height$(q/p') <$ height(q/p) and we have a contradiction, proving that height$(q/p) = 1$. This now contradicts the hypothesis.

In [M1] it is shown that if $P \subseteq Q$ has going down for any Q and any height 1 prime P, then the Noetherian domain R has going down. This can be combined with our previous ideas. To illustrate this in an easy setting, we will consider a catenary Noetherian domain.

PROPOSITION 9.16. Let R be a catenary Noetherian domain and let n be less than the height of any G-ideal of R. Suppose that $P \subset Q$ has going down whenever Q is a G-ideal and height $P = n$. Then R has going down.

Proof: We first inductively reduce to the case that $n = 1$. For this, suppose the hypothesis holds for n and let p be a prime of height $n-1$. Furthermore let Q be a G-ideal containing p. Since R is catenary and height $Q > n$, height$(Q/P) > 1$. Thus $W = \{P \in \text{spec } R \mid p \subset P \subset Q \text{ and height}(P/p) = 1\}$ is infinite. By Corollary 9.5, for some $P \in W$, $p \subset P$ has going down. Since height $P =$ height $p + 1 = n$, by induction, $P \subset Q$ has going down. Thus $p \subset Q$ clearly has going down.

We now have $P \subset Q$ has going down whenever Q is a G-ideal and height $P = 1$. The technique in the first paragraph of the proof of Proposition 9.15 allows us to conclude that $P \subset Q$ has going down for any height 1 P and any Q. The result in [M1] now shows that R has going down.

We close with an example of primes $p \subset p_1 \subset p_2$ in a Noetherian domain for which $p \subseteq p_1$ does not satisfy going down, while $p \subset p_2$ does satisfy it. Thus the chain of ideals $p = I_0 \subset I_1 \subset I_2 \subset \ldots \subset I_n$ discussed in Proposition 9.9 has $n \geq 2$.

EXAMPLE. Let $0 \subset P_1 \subset P_2$ be primes in a Noetherian domain R, and suppose that in the integral closure $\bar{R}$, at least two primes, Q_1 and Q_2, lie over P_1. Let $c \in \bar{R}$ with $c \in Q_1 - Q_2$ and with c in every prime of $\bar{R}$ which lies over P_2. Now in $\bar{R}[X]$, $q = (X-c)\bar{R}[X]$ is prime. Let $p = q \cap R[X]$. If $S = R - \{0\}$ and $K = R_S$, then $p_S = q_S = (X-c)K[X]$, and we easily see that q is the unique prime of $\bar{R}[X]$ lying over p.

Now $c \in Q_1$, so $q \subset (Q_1, X)\bar{R}[X]$. Contracting to R, we have $p \subset (P_1, X)R[X]$ We claim going down fails here. This is obvious, since $(Q_2, X)\bar{R}[X]$ lies over $(P_1, X)R[X]$ and $q \not\subset (Q_2, X)\bar{R}[X]$ since $c \notin Q_2$. On the other hand, $p \subset (P_2, X)R[X]$ does satisfy going down, since every prime of $\bar{R}[X]$ lying over $(P_2, X)R[X]$ has form $(Q,X)\bar{R}[X]$, with Q a prime of $\bar{R}$ lying over P_2. By choice of c, $c \in Q$, so that $q \subset (Q,X)\bar{R}[X]$. Thus we have $p \subset (P_1, X)R[X] \subset (P_2, X)R[X]$ with $p \subset (P_1, X)R[X]$ failing going down, while $p \subset (P_2, X)R[X]$ satisfies going down.

CHAPTER X: Ideal Transforms

In Chapter 6, we called ideals I and J projectively equivalent, $I \sim J$, if for some positive integers n and m $\overline{I^n} = \overline{J^m}$. Obviously, in a Noetherian ring, $I \sim J$ implies that $\overline{A}^*(I) = \overline{A}^*(J)$, so that by Proposition 3.17 we see that $\overline{A}^*(I) \subseteq \cap\, \overline{A}^*(J)$, $J \sim I$. We give an example showing that this inclusion may be proper.

EXAMPLE. Let (R,M) be a 2-dimensional local domain which is quasi-unmixed but not unmixed. That is, every minimal prime of R^* has depth 2, but there is a prime divisor of zero, Q, in R^* with depth $Q = 1$. (Such an R was constructed in [F-R]. A more general construction was recently given in [B-R].). Let $0 \neq I = aR$ be a principal ideal. If $J \sim I$, then since $Q \cap R = 0$, we have $JR^* \not\subseteq Q$. Thus M^* is minimal over $JR^* + Q$, and by Proposition 1.14, $M^* \in A^*(JR^*)$. Therefore $M \in A^*(J)$ for all $J \sim I$. On the other hand, Proposition 3.19 shows that $M \notin \overline{A}^*(I)$ since R^* does not have a depth 1 minimal prime.

In general, we do not understand the differences between $\overline{A}^*(I)$ and $\cap A^*(J)$, $J \sim I$ very well, except in the case that I is principal. We explore that case in this chapter, and see that it has connections to known results concerning ideal transforms.

DEFINITION. Let R be a Noetherian ring with total quotient ring $Q(R)$. Let I be a regular ideal of R. The ideal transform of I is $T(I) = \{y \in Q(R) \mid yI^n \subseteq R$ for some $n\}$.

LEMMA 10.1. Let I be a regular ideal of the Noetherian ring R. Then

i) If $x \in I$ is a regular element, $T(I) \subseteq R_x$

ii) If $I = (x_1, \ldots, x_n)$ with each x_i regular, then $T(I) = R_{x_1} \cap \ldots \cap R_{x_n}$. (Note: such x_i always exist.)

iii) If $R \subseteq \hat{R}$ with $\hat{R}$ a flat Noetherian extension of R, then $T(I\hat{R}) = T(I) \otimes_R \hat{R}$. In particular, if S is a multiplicatively closed set in R, then $T(I)_S = T(I_S)$.

iv) If $P \in \operatorname{spec} R$ and $I \not\subseteq P$, then $T(I)$ contains a unique prime Q lying over P, and $T(I)_Q = R_P$.

v) If $x \in I$ is a regular element, and if y is in the total quotient ring of R and satisfies $Iy \subseteq xT(I)$, then $y \in xT(I)$.

vi) I does not consist of zero divisors modulo $xT(I) \cap R$ for any regular $x \in I$.

Proof: These are fairly simple. We prove only (v) and (vi). For (v), since Iy is finitely generated, for some large n we have $I^{n+1}y \subseteq xR$. As $y/x \in Q(R)$, and $I^{n+1}(y/x) \subseteq R$, we have $y/x \in T(I)$. This proves (v), and (vi) follows easily.

In [Ni], Nishimura investigates when $T(I)$ is either an integral extension of R, or a finite R-module. We reproduce and extend Nishimura's work, first looking at the integral case. The next lemma simplifies matters. Recall that $z(P)$ is the minimal depth of a minimal prime in $(R_P)^*$.

DEFINITION. If R is a Noetherian ring with total quotient ring $Q(R)$, then $R^{[1]} = \{y \in Q(R) \mid (R : y)_R \not\subseteq P$ for all primes P with $z(P) = 1\}$.

LEMMA 10.2. $R^{[1]} \subseteq \overline{R}$.

Proof: Suppose $a/b \in Q(R) - \overline{R}$. Then $(\overline{R} : a/b)_R$ is a proper ideal of R. If P is a prime divisor of this ideal, then for some $c \in R$ we have $P = (b\overline{R} : ac) = (\overline{bR} : ac)$ since $b\overline{R} \cap R = \overline{bR}$. Thus $z(P) = 1$. As $(R : a/b) \subseteq (\overline{R} : a/b) \subseteq P$, we have $a/b \notin R^{[1]}$.

PROPOSITION 10.3 Let (R,M) be a local ring with M regular. The following are equivalent.

i) $T(M) \subseteq \overline{R}$.

ii) $T(M) \subseteq R^{[1]}$.

iii) R^* does not contain a depth 1 minimal prime.

Proof: (ii) $\Rightarrow$ (i): By Lemma 10.2.

(iii) $\Rightarrow$ (ii): Clearly $T(M) = \{y \in Q(R) \mid M \subseteq \operatorname{rad}(R : y)\}$. By (iii), $z(M) \neq 1$, and we see that $T(M) \subseteq R^{[1]}$.

i) $\Rightarrow$ iii): We will prove the contrapositive. Thus assume that (iii) fails. Since $T(M^*) = T(MR^*) = T(M) \otimes_R R^*$, to show (i) fails it will suffice to show that $T(M^*) \not\subseteq \overline{R^*}$. Thus we may assume that R is complete. As (iii) fails, there is a depth 1 minimal prime Q of R. Let $x \in M$ be regular, so that M is minimal over $(x)+Q$. Suppose $M^k \subseteq (x)+Q$. Now for some $s \in R-Q$ and $n \geq 1$, $sQ^n = 0$. By Lemma 3.11, we may also assume $s \notin \overline{(x^n)}$ by increasing n if necessary. Now $M^{2nk} \subseteq ((x)+Q)^{2n} \subseteq (x^n)+Q^n$. Thus $M^{2nk}s \subseteq (x^n)$. Therefore $s/x^n \in T(M)$, but $s/x^n \notin \overline{R}$ since $s \notin \overline{(x^n)} = x^n\overline{R} \cap R$, and (i) fails.

COROLLARY 10.4. Let I be a regular ideal in a Noetherian ring. The following are equivalent:

i) $T(I) \subseteq \overline{R}$.

ii) $T(I) \subseteq R^{[1]}$.

iii) $z(P) > 1$ for each prime P containing I.

iv) $z(P) > 1$ for each $P \in \overline{A}^*(I)$.

Proof: (i) $\Rightarrow$ (iv): If $P \in \overline{A}^*(I)$, then $I \subseteq P$ so that $T(P) \subseteq T(I) \subseteq \overline{R}$ using (i). Therefore $T(P_P) \subseteq \overline{R_P}$ and Proposition 10.3 gives $z(P) > 1$.

(iv) $\Rightarrow$ (iii). We treat the contrapositive. Thus suppose that $I \subseteq P$ and $z(P) \leq 1$. Since I is regular, clearly $z(P) \neq 0$, and so $z(P) = 1$. Let q be a depth 1 minimal prime in $(R_P)^*$. As $(P_P)^*$ is minimal over $IR_P^* + q$, by Proposition 3.18 we see that $P \in \overline{A}^*(I)$. As $z(P) \neq 1$, (iv) fails. This shows that (iv) $\Rightarrow$ (iii).

(iii) $\Rightarrow$ (ii): Since $T(I) = \{y \in Q(R) \mid I \in \text{rad}(R:y)\}$, this is obvious from the definition of $R^{[1]}$.

(ii) $\Rightarrow$ (i): Immediate from Lemma 10.2.

COROLLARY 10.5. Let R be a Noetherian ring Then $R^{[1]} = \cup T(I)$ over all regular ideals I such that $T(I) \subseteq \overline{R}$.

Proof: One containment is given immediately by the preceding corollary. If $y \in R^{[1]}$, let $I = (R:y)$ which is a regular ideal since $y \in Q(R)$. As $y \in R^{[1]}$, if P is a prime containing I, then $z(P) > 1$. Now the preceding corollary tells us that $T(I) \subseteq R^{[1]}$. As $y \in T(I)$, we are done.

COROLLARY 10.6. Let (R,M) be a local ring and let $n = \min\{\text{depth } q \mid q \text{ is a minimal prime in } R^*\}$. If I is a regular ideal with depth $I \leq n-2$, then $T(I) \subseteq \bar{R}$.

Proof: By Corollary 10.4 we must show that $z(P) > 1$ for every prime P containing I. Suppose to the contrary that $z(P) = 1$ ($z(P) = 0$ being impossible since I is regular). Now $(R_P)^*$ has a depth 1 minimal prime. Also $(R/P)^*$ obviously has a minimal prime of depth equal to depth P. Thus by a result proved in the Appendix, R^* has a minimal prime of depth equal to depth $P + 1 \leq$ depth $I + 1 \leq n-2+1 = n-1$, a contradiction.

We remark that for a Noetherian domain R, $R^{[1]} = \cap R_P$, $z(P) = 1$. Thus $R^{[1]}$ is in concept similar to the well known extension $R^{(1)} = \cap R_P$, height $P = 1$. It is interesting that for (R,M) local, these two extensions differ by only a finite amount (in some sense) as we now show

PROPOSITION 10.7. Let (R,M) be a local ring. The set $\{P \in \operatorname{spec} R \mid z(P) = 1$ but height $P > 1\}$ is finite.

Proof: Suppose that P is in the given set. By Proposition 5.6, $\operatorname{gr}^* P = 1$. If a is an asymptotic sequence from P, then $P \in \bar{A}^*(aR)$. By Proposition 3.18, there are primes $q^* \subseteq P^*$ or R^* with q^* minimal, $P^*/q^* \in \bar{A}^*(aR^*+q^*/q^*)$ and $P^* \cap R = P$. By Lemma 3.14, since R^*/q^* satisfies the Altitude Formula, we have height $P^*/q^* = 1$. However height $P^* \geq$ height $P > 1$. Thus P^* must contain some minimal prime besides q^*. If J^* is the intersection of the other minimal primes, then $J^* \subseteq P^*$. Since $J^* \not\subseteq q^*$, only finitely many primes can have height $P^*/q^* = 1$ and height $P^* > 1$. As this is true for each of the finitely many minimal q^*, we are done.

We now begin to consider when $T(I)$ is a finite R-module. Previously, the following was known ((ii) $\Leftrightarrow$ (iii) due to Nishimura, (i) $\Leftrightarrow$ (iii) due to Ratliff [R4]).

THEOREM. Let (R,M) be a local ring with M regular. The following are equivalent.

i) R^* has a depth 1 prime divisor of zero.

ii) $T(M)$ is an infinite R-module.

iii) There is an integer n such that $M \in \mathrm{Ass}(R/J)$ for all ideals $J \subseteq M^n$.

Our contribution (this being work done with Katz) will be to add the following two statements to this list of equivalences.

iv) $M \in A^*(J)$ for any regular ideal $J \subseteq P$.

v) There is a regular $x \in M$ with $M \in \cap A^*(J)$, $J \sim xR$.

LEMMA 10.8. Let I be an ideal of the Noetherian ring R and let $x \in I$ be a regular element. Then $T(I)$ is a finite R-module if and only if $T(I_P)$ is a finite R_P-module for each $P \in \mathrm{Ass}(R/xR)$ with $I \subseteq P$.

Proof: One direction is immediate. Thus suppose that $T(I_P)$ is a finite R_P-module for each $P \in \mathrm{Ass}(R/xR)$ with $I \subseteq P$. For each such P, there is an n with $x^n T(I_P) \cap R_P \subseteq xR_P$ (Artin-Rees Lemma). If m is the maximum of those finitely many n, then $x^m T(I_P) \cap R_P \subseteq xR$ for all $P \in \mathrm{Ass}(R/xR)$ (since if $I \not\subseteq P$, $T(I_P) = R_P$) from which it follows that $x^m T(I) \cap R \subseteq xR$. We claim $T(I) \subseteq Rx^{-m}$. Since $x \in I$, $T(I) \subseteq R_x$. For $a \in T(I)$ choose k minimal with $x^k a \in R$. If $k > m$ then $x^k a \in x^m T(I) \cap R \subseteq xR$ which says $x^{k-1} a \in R$, a contradiction Therefore $k \leq m$, showing that $T(I) \subseteq Rx^{-m}$. Thus $T(I)$ is a finite R-module.

The thrust of the next proposition is that to understand the finiteness of $T(I)$, it is enough to understand $T(P_P)$, for primes P containing I. In Proposition 10.10, our key result, we will examine $T(P_P)$.

PROPOSITION 10.9. (Nishimura) Let I be an ideal in the Noetherian ring R, and let $x \in I$ be a regular element. The following are equivalent.

i) $T(I)$ is a finite R-module.

ii) $T(P_P)$ is a finite R_P-module for all primes P containing I.

iii) $T(P_P)$ is a finite R_P-module for primes $P \in \mathrm{Ass}(R/xR)$ with $I \subseteq P$.

Proof: (i) $\Rightarrow$ (ii): This is trivial since $I \subseteq P$ implies $T(P) \subseteq T(I)$.

(ii) $\Rightarrow$ (iii): Trivial.

(iii) $\Rightarrow$ (i): We make use of Lemma 10.8. Therefore let $P \in \mathrm{Ass}(R/xR)$ with $I \subseteq P$. By (iii) we have that $T(P_P)$ is a finite R_P-module, and our goal is to show that $T(I_P)$ is a finite R_P-module. Suppose this is false for some such P. We may assume that P is a minimal counterexample. Letting $A = T(P_P)$, we claim that $T(IA)$ is a finite A-module. In order to see this, we again use Lemma 10.8. Therefore let $Q \in \mathrm{Ass}(A/xA)$ with $IA \subseteq Q$, and let $Q \cap R_P = q_P$ ($q \in \mathrm{spec}\ R$). Suppose that $Q = (xA : y)$, $y \in A - xA$. Then $q_P y \subseteq xA = xT(P_P)$. Since $y \notin xA$, Lemma 10.1(v) shows that $q_P \neq P_P$. Now Lemma 10.1(iv) shows that $A_Q = (R_P)_{q_P} = R_q$, so that $T(IA_Q) = T(I_q)$. As $Q \in \mathrm{Ass}(A/xA)$ and $IA \subseteq Q$, we see that $q \in \mathrm{Ass}(R/xR)$ and $I \subseteq q$. By the minimality of P, we have that $T(I_q) = T(IA_Q)$ is a finite $R_q = A_Q$ - module. By Lemma 10.8, our claim that $T(IA)$ is a finite A-module is now proved. As (iii) says that A is a finite R_P-module, we deduce that $T(IA)$ is a finite R_P-module. Finally, since $R_P \subseteq A \subseteq Q(R_P)$, obviously $T(I_P) \subseteq T(IA)$. This contradicts our assumption that $T(I_P)$ failed to be a finite R_P-module.

LEMMA 10.10. Let I be an ideal in a local ring (R,M). If R is complete in the M-adic topology, then it is complete in the I-adic topology.

Proof: Let $\{x_n\}$ be an I-Cauchy sequence. We may assume that $x_{n+1} - x_n \in I^n$. Therefore if $I^n = (a_1, \ldots, a_m)$ then for $k \geq 1$, $x_{n+k} - x_n = (x_{n+k} - x_{n+k-1}) + \ldots + (x_{n+1} - x_n) = a_1(\Sigma r_{1j}) + \ldots + a_m(\Sigma r_{mj})$ with $j = 0,1,\ldots,k-1$ and $r_{ij} \in I^j$ $i = 1,\ldots,m$. Being I-Cauchy, $\{x_n\}$ is also M-Cauchy and so $x_n \to x \in R$ in the M-adic topology. Taking M-adic limits, obviously $x - x_n = \lim_{k\to\infty}(x_{n+k} - x_n) = a_1(\lim_{k\to\infty}\Sigma r_{1j}) + \ldots + a_m(\lim_{k\to\infty}\Sigma r_{mj})$. We note that $\lim_{k\to\infty}\Sigma r_{ij}$ exists in R since $r_{ij} \in I^j \subseteq M^j$. Thus clearly $x - x_n \in I^n$ showing that $x_n \to x$ in the I-adic topology.

PROPOSITION 10.11. Let P be a regular prime in a Noetherian ring R. The following are equivalent.

i) $T(P_P)$ is an infinite R_P-module.

ii) $(R_P)^*$ contains a depth 1 prime divisor of zero.

iii) There is an $m \geq 1$ such that for any regular ideal $J \subseteq P^{(m)}$, $P \in \mathrm{Ass}(R/J)$.

iv) $P \in A^*(J)$ for any regular ideal $J \subseteq P$.

v) There is a regular element $x \in P$ with $P \in \cap A^*(J)$, $J \sim xR$.

vi) Either height $P = 1$ or there is a regular element $x \in P$ with $P \in A^*(x^k T(P) \cap R)$ for all large k.

Proof: (i) $\Rightarrow$ (ii): For this, we may assume that R is local at P. Letting R^* be the P-adic completion, we have R^* a faithfully flat extension of R. Since $T(P)$ is an infinite R-module, $T(P^*) = T(PR^*) = T(P) \otimes R^*$ is an infinite R^*-module. Now by Lemma 10.10, if $x \in P$ is a regular element, then R^* is also xR^*-adic complete. By [Ma], $T(P^*)/xT(P^*)$ is a finite R^* module. Thus by [Mt, Lemma p. 212] we must have $\cap x^n T(P^*) \neq 0$.

Therefore we may select $0 \neq y \in (\cap x^n T(P^*)) \cap R^*$. A well known corollary of the Artin-Rees Lemma gives us that for some k and large n, $(x^n : y)_{R^*} \subseteq (0 : y)_{R^*} + x^{n-k}R^*$. As $y \in x^n T(P^*)$, we see that $(x^n : y)_{R^*}$ is P^* primary. Thus $(0 : y)_{R^*} + x^{n-k}R^*$ is P^* primary. Since $y \neq 0$, this obviously shows that R^* has a depth 1 prime divisor of zero. Thus (i) $\Rightarrow$ (ii).

(ii) $\Rightarrow$ (iii): For this we again may localize at P (so that $P^{(n)} = P^n$) and pass to the completion $(R_P)^*$. Then Lemma 1.13 gives that (iii) holds.

(iii) $\Rightarrow$ (iv) $\Rightarrow$ (v) are obvious.

(v) $\Rightarrow$ (i): Suppose that (v) holds but that $T(P_P)$ is a finite R_P-module. We will derive a contradiction. We have $T(P_P) \subseteq \overline{R_P}$ (the integral closure of R_P in its total quotient ring). This easily implies that $T(P_P) = (T(P) \cap \overline{R})_P$. Now for any $k \geq 1$, $x^k R \subseteq x^k(T(P) \cap \overline{R}) \cap R \subseteq x^k\overline{R} \cap R = \overline{x^k R}$, so that $x^k(T(P) \cap \overline{R}) \cap R \sim xR$. By (v), $P \in A^*(x^k(T(P) \cap \overline{R}) \cap R)$. Localizing, we have $P_P \in A^*(x^k T(P_P) \cap R_P)$. As $T(P_P)$ is a finite R_P-module (our assumption), it is an easy consequence of the Artin-Rees Lemma that for large k and all $n \geq 1$, $(x^k T(P_P) \cap R_P)^n = x^{nk} T(P_P) \cap R_P$. Since $P_P \in A^*(x^k T(P_P) \cap R_P)$, for large n we have P_P a prime divisor of $x^{nk} T(P_P) \cap R_P$. This contradicts Lemma 10.1(vi). Thus (v) $\Rightarrow$ (i).

(iv) ⇒ (vi): Suppose (iv) holds. If $x \in P$ is a regular element and if $x^kT(P) \cap R \subseteq P$, then (iv) implies $P \in A^*(x^kT(P) \cap R)$ and we are done. Therefore we need only worry about the case that $x^kT(P) \cap R \not\subseteq P$. Then $x^kT(P_P) \cap R_P = R_P$. Thus $x^{-k} \in T(P_P)$ so that for some m, $P_P^m \subseteq x^kR_P$. This shows that height $P = 1$, proving (vi).

(vi) ⇒ (i): If height $P = 1$ then clearly (iv) holds, so (i) holds. Suppose that $P \in A^*(x^kT(P) \cap R)$ for large k. Then $P_P \in A^*(x^kT(P_P) \cap R_P)$ and the argument employed in proving (v) ⇒ (i) works again.

Question: Is $P \in A^*(\overline{x^kR})$, $x \in P$ regular, k large, equivalent to the conditions in the preceding proposition?

We may now add another equivalent statement to the list in Proposition 10.9, analogous to statement (iv) of Corollary 10.4.

COROLLARY 10.12. Let I be a regular ideal in a Noetherian ring. The following are equivalent.

i) $T(I)$ is a finite R-module.

ii) $T(P_P)$ is a finite R_P-module for all $P \in A^*(I)$.

Proof: (i) ⇒ (ii) is by Proposition 10.9. Suppose now that (i) fails. Then by Proposition 10.9 there is some prime P containing I such that $T(P_P)$ is an infinite R_P-module. By Proposition 10.11, $P \in A^*(I)$, showing that (ii) fails.

LEMMA 10.13. Let P be a regular prime in a Noetherian ring R. Then $T(P_P)$ is an infinite R_P-module if and only if for some $Q \in \mathrm{Ass}(R)$ with $Q \subset P$, $T(P'_{P'})$ is an infinite $R'_{P'}$-module, the primes denoting modulo Q.

Proof: Suppose $T(P_P)$ is an infinite R_P-module. By Proposition 10.11, $(R_P)^*$ has a depth 1 prime divisor of zero, say Q^*. Let $Q^* \cap R_P = Q_P$. Then $Q_P \in \mathrm{Ass}(R_P)$ (so $Q \subset P$ and $Q \in \mathrm{Ass}\, R$). Now $(R'_{P'})^* \approx (R_P/Q_P)^* \approx (R_P)^*/Q_P(R_P)^*$ and $Q^*/Q_P(R_P)^*$ is a depth 1 prime divisor of zero in this ring. Thus $T(P'_{P'})$ is an infinite $R'_{P'}$-module. The converse is similar.

LEMMA 10.14. Let $R \subseteq \hat{R}$ be a flat extension of Noetherian rings. Let P be a regular prime of R and let $\hat{P} \in \operatorname{spec} \hat{R}$ with $\hat{P} \cap R = P$. If $T(\hat{P}_{\hat{P}})$ is an infinite $\hat{R}_{\hat{P}}$-module, then $T(P_P)$ is an infinite R_P-module. Furthermore, if $\hat{R}$ is a faithfully flat extension and if $T(P_P)$ is an infinite R_P-module then for some $\hat{P} \in \operatorname{spec} \hat{R}$ lying over P, $T(\hat{P}_{\hat{P}})$ is an infinite $\hat{R}_{\hat{P}}$-module.

Proof: For the first assertion, we have that $\hat{P}$ satisfies (iv) of Proposition 10.11 Now if $J \subseteq P$ with J a regular ideal of R, we see that $\hat{P} \in A^*(J\hat{R})$. Thus $P \in A^*(J)$ and we use Proposition 10.11.

For the second assertion, we let $S = R - P$ and note that $(\hat{R})_S$ is a faithfully flat R_S-module. Since $T(P_S)$ is an infinite R_S-module, $T(P\hat{R}_S) = T(P_S) \otimes_{R_S} \hat{R}_S$ is an infinite $\hat{R}_S$-module. By Proposition 10.9, there is a prime $\hat{P}_S \in \operatorname{spec} \hat{R}_S$ with $P\hat{R}_S \subseteq \hat{P}_S$ such that $T(\hat{P}_{\hat{P}})$ is an infinite $\hat{R}_{\hat{P}}$-module (using $(\hat{R}_S)_{\hat{P}_S} = \hat{R}_{\hat{P}}$). Clearly $\hat{P} \cap R = P$ and we are done.

PROPOSITION 10.15. Let P be a regular prime in a local ring (R,M) with completion R^*. Then $T(P_P)$ is an infinite R_P-module if and only if there are primes $Q^* \subset P^*$ in R^* with $Q^* \in \operatorname{Ass} R^*$, $P^* \cap R = P$ and height $P^*/Q^* = 1$.

Proof: Suppose first that such $Q^* \subset P^*$ exist. Let J be a regular ideal of R with $J \subseteq P$. Clearly P^* is minimal over $JR^* + Q^*$, so by Proposition 1.14, $P^* \in A^*(JR^*)$. Thus $P \in A^*(J)$ and so we use Proposition 10.11. Conversely, suppose that $T(P_P)$ is an infinite R_P-module. By Lemma 10.14, there is a $P^* \in \operatorname{spec} R^*$ with $P^* \cap R = P$ and $T(P^*_{P^*})$ an infinite $R^*_{P^*}$ module. By Lemma 10.13 there is a $Q^* \in \operatorname{Ass}(R^*)$ with $Q^* \subset P^*$ and (with primes denoting modulo Q^*) $T(P^{*\prime}_{P^{*\prime}})$ an infinite $R^{*\prime}_{P^{*\prime}}$-module. By Proposition 10.11, $P^{*\prime} \in A^*(J)$ for every ideal $J \neq 0$ of $R^{*\prime}$ (since $R^{*\prime}$ is a domain). Let $x \neq 0$ be any element of $P^{*\prime}$. Since $R^{*\prime}$ is analytically unramified, by [SO], for large m, $J = \overline{(x^m R^{*\prime})}$ is normal, $(J^n = \overline{J^n}$ all $n)$. As $P^{*\prime} \in A^*(J)$, for large n, $P^{*\prime}$ is a prime divisor of $J^n = \overline{J^n}$ $= \overline{(x^{mn} R^{*\prime})}$. Thus $P^{*\prime} \in \overline{A}^*(xR^{*\prime})$. However as R^* is complete, $R^{*\prime}$ satisfies the altitude formula so that Lemma 3.14 shows that height $P^{*\prime} = 1$. Thus height $P^*/Q^* = 1$ as desired.

COROLLARY 10.16. Let P be a regular prime in a local ring (R,M). Then $(R_P)^*$ contains a depth 1 prime divisor of zero if and only if there exist primes $Q^* \subset P^*$ of R^* with $P^* \cap R = P$, $Q^* \in \mathrm{Ass}(R^*)$ and height $P^*/Q^* = 1$.

Proof: Immediate from Proposition 10.15 and 10.11.

We do not know of a direct proof of the preceding. For the analogous result for minimal primes, we do not need ideal transforms (hence we do not need P regular) as we now show.

PROPOSITION 10.17. Let P be a non-minimal prime in a local ring (R,M). Then $(R_P)^*$ contains a depth 1 minimal prime if and only if there exist primes $Q^* \subset P^*$ in R^* with $P^* \cap R = P$, Q^* minimal, and height $P^*/Q^* = 1$.

Proof: Suppose such $Q^* \subset P^*$ exist. Let $x \in P$ with x in no minimal prime of R. Then $x \notin Q^*$ and so P^* is minimal over $xR^* + Q^*$, so that $P^* \in \overline{A}^*(xR^*)$ by Corollary 3.13. Proposition 3.18 gives $P \in \overline{A}^*(xR)$, and so $P_P \in \overline{A}^*(xR_P)$. Thus $\mathrm{gr}^* P_P = 1$, and so Proposition 5.6 shows that $z(P) = z(P_P) = 1$ as desired. Conversely suppose that $z(P) = 1$. With x as above, Corollary 5.12 shows that $P \in \overline{A}^*(xR)$ (since x in no minimal prime gives $\mathrm{gr}^* xR > 0$). By Proposition 3.18 used twice, there are primes $Q^* \subset P^*$ of R^* with $P^* \cap R = P$, Q^* minimal, and $P^*/Q^* \in \overline{A}^*(xR^* + Q^*/Q^*)$. As R^*/Q^* satisfies the altitude formula, Lemma 3.14 gives that height $P^*/Q^* = 1$.

COROLLARY 10.18. Let P be a regular prime in a local ring (R,M). Then $T(P_P) \not\subseteq \overline{R}_P$ if and only if there are primes $Q^* \subset P^*$ in R^* with $P^* \cap R = P$, Q^* minimal and height $P^*/Q^* = 1$.

Proof: Immediate from Propositions 10.17 and 10.3.

PROPOSITION 10.19. Let I be a regular ideal in a local ring (R,M). Let $n = \min\{\mathrm{depth}\ q^* \mid q^* \text{ is a minimal prime in } R^*\}$. Let $m = \min\{\mathrm{depth}\ Q^* \mid Q^* \in \mathrm{Ass}(R^*)\}$.

a) If $n = 1$ then $M \in \overline{A}^*(I)$ and $T(I) \not\subseteq \overline{R}$.

b) If $n = 2$ then either $M \in \overline{A}^*(I)$ or $T(I) \not\subseteq \overline{R}$.

c) If $m=1$ then $M \in A^*(I)$ and $T(I)$ is an infinite R-module

d) If $m=2$ then either $M \in A^*(I)$ or $T(I)$ is an infinite R-module.

Proof: (a): Since $n=1$, $z(M)=1$. Proposition 5.6 and Corollary 5.12 show $M \in \overline{A}^*(I)$. Corollary 10.4 shows $T(I) \not\subseteq \overline{R}$.

(b): Let q^* be a depth 2 minimal prime in R^*. Suppose that $M \notin \overline{A}^*(I)$. Then $M^* \notin \overline{A}^*(IR^*)$ by Proposition 3.18, so that Corollary 3.13 shows that M^* is not minimal over IR^*+q^*. Thus there is a prime p^* with $IR^*+q^* \subseteq p^* \subset M^*$. As depth $q^*=2$, height $p^*/q^*=1$. Letting $p=p^* \cap R$, Corollary 10.18 yields $T(p_p) \not\subseteq \overline{R}_p$. By Proposition 10.3, $z(p)=1$. As $I \subseteq p$, Corollary 10.4 gives $T(I) \not\subseteq \overline{R}$.

(c): Use Propositions 10.11 and 10.9.

(d): Let $Q^* \in \mathrm{Ass}(R^*)$ have depth 2. If $M \notin A^*(I)$ then $M^* \notin A^*(IR^*)$ and Proposition 1.14 shows M^* is not minimal over IR^*+Q^*. Let P^* be prime with $IR^*+Q^* \subseteq P^* \subset M^*$. Depth $Q^*=2$ implies height $P^*/Q^*=1$. Letting $P=P^* \cap R$, we use Propositions 10.15 and 10.9 to see that $T(I)$ is an infinite R-module.

We proceed to prove analogues of Corollaries 10.6 and 10.5.

COROLLARY 10.20. Let (R,M) be a local ring and let $n=\min\{\text{depth } Q^* \mid Q^* \in \mathrm{Ass}(R^*)\}$. If I is a regular ideal of I with depth $I \leq n-2$, then $T(I)$ is a finite R-module.

Proof: If false, then by Propositions 10.9 and 10.15 there are primes $Q^* \subset P^*$ in R^* with $Q^* \in \mathrm{Ass}(R^*)$, height $P^*/Q^*=1$ and $I \subseteq P^* \cap R$. As R^*/Q^* is catenary, depth $Q^* = \text{depth } P^*+1 \leq \text{depth } IR^*+1 = \text{depth } I+1 \leq n-1$, contradicting the definition of n.

$R^{[1]}$ was defined in terms of primes with $z(P)=1$, that is, with $(R_P)^*$ having a depth 1 minimal prime. We now define $R^{\langle 1 \rangle}$ analogously.

DEFINITION. Let R be a Noetherian ring with total quotient ring $Q(R)$. $R^{\langle 1\rangle} = \{y \in Q(R) \mid (R: y) \not\subseteq P$ for all primes P such that $(R_P)^*$ contains a depth 1 prime divisor of zero$\}$.

PROPOSITION 10.21. Let I be a regular ideal in a Noetherian ring R. Then $T(I) \subseteq R^{\langle 1\rangle}$ if and only if $T(I)$ is a finite R-module. Also $R^{\langle 1\rangle} = \cup\, T(I)$ over regular ideals I such that $T(I)$ is a finite R-module.

Proof: Suppose that $T(I)$ is a finite R-module. By Propositions 10.9 ((i)$\Rightarrow$(ii)) and 10.11 ((ii)$\Rightarrow$(i)), for any prime P containing I, $(R_P)^*$ does not have a depth 1 prime divisor of zero. As $T(I) = \{y \in Q(R) \mid I \subseteq \mathrm{rad}(R: y)\}$, clearly $T(I) \subseteq R^{\langle 1\rangle}$. For the converse, suppose that $T(I)$ is an infinite R-module. Then Proposition 10.9 ((ii)$\Rightarrow$(i)) and 10.11 ((i)$\Rightarrow$(v)) show there is a regular element $x \in P \supseteq I$ with $P \in A^*(xR)$. Thus for some integer n and $c \in R$, $P = (x^n: c)$. Now $y = c/x^n \in Q(R)$, and $Iy \subseteq Py \subseteq R$, so that $y \in T(I)$. On the other hand, since $(R: y) = P$ and since $(R_P)^*$ has a depth 1 prime divisor of zero (Proposition 10.11), we have $y \notin R^{\langle 1\rangle}$. Thus $T(I) \not\subseteq R^{\langle 1\rangle}$.

Clearly $\cup\, T(I)$, $T(I)$ finite, is contained in $R^{\langle 1\rangle}$. For the reverse inclusion say $z \in R^{\langle 1\rangle}$ and let $I = (R: z)$. As $z \in Q(R)$, I is regular. Since $z \in R^{\langle 1\rangle}$, if P is a prime containing I, then $(R_P)^*$ has no depth 1 prime divisor of zero. Propositions 10.9 and 10.11 now show that $T(I)$ is a finite R-module. Thus $T(I) \subseteq R^{\langle 1\rangle}$. As $z \in T(I)$, we are done.

Remark: In [B3], Brodmann shows that in a local ring (R,M), if the characteristic of R/M is either 0 or a regular element of R, and if $n = \min\{\text{depth } Q^* \mid Q^* \in \mathrm{Ass}(R^*)\}$, then $\cup\{T(I) \mid \text{depth } I \leq n-2\}$ is a finite R-module. This is stronger than our Corollary 10.20. We ask a question, which if true, is stronger still.

Question: If (R,M) is a local ring, is $R^{\langle 1\rangle}$ a finite R-module?

Remark: In [Sc], the following result is reported. Let (R,M) be a local ring. The following are equivalent: (a) There is a $Q^* \in \mathrm{Ass}(R^*)$ with depth $Q^* \leq n$,

(b) There is an integer k with $M \in \mathrm{Ass}(R/I)$ for every ideal $I \subseteq M^k$ with height $I \geq 0$. This is an extension of our Proposition 10.11 ((ii) $\Leftrightarrow$ (iv)). We do not know if Proposition 10.11 ((ii) $\Leftrightarrow$ (v)) can also be extended. We now present a major question related to this.

Question: Suppose the sequence $x_1, \ldots, x_n$ in a Noetherian ring R is called a strong asymptotic sequence if $(x_1, \ldots, x_n) \neq R$ and for $i = 1, \ldots, n$, $x_i \notin \cap A^*(J)$, $J \sim (x_1, \ldots, x_{i-1})$. Our question is, if (R,M) is a local ring, will $x_1, \ldots, x_n$ be a strong asymptotic sequence if and only if $\mathrm{height}((x_1, \ldots, x_n)R^* + Q^*/Q^*) = n$ for all $Q^* \in \mathrm{Ass}(R^*)$? This is true if $n = 1$ or $n = 2$. The case $n = 1$ is trivial since $\cup A^*(J)$, $J \sim 0$ equals $\mathrm{Ass}(R)$. For $n = 2$, let x_1, x_2 be a strong asymptotic sequence in (R,M). Suppose also that $Q^* \in \mathrm{Ass}(R^*)$ and $\mathrm{height}((x_1, x_2)R^* + Q^*/Q^*) < 2$. We will derive a contradiction. Since x_1 is regular, this height is 1. Suppose P^* is prime in R^* with $(x_1, x_2)R^* + Q^* \subseteq P^*$ and height $P^*/Q^* = 1$. If $P = P^* \cap R$, Corollary 10.16 shows that $(R_P)^*$ contains a depth 1 prime divisor of zero, so that Proposition 10.11 ((ii) $\Rightarrow$ (iv)) gives $P \in \cap A^*(J)$, $J \sim x_1R$. Since $x_2 \in P^* \cap R = P$, we have contradicted that x_1, x_2 is a strong asymptotic sequence. Conversely, suppose $\mathrm{height}((x_1, x_2)R^* + Q^*/Q^*) = 2$ for every $Q^* \in \mathrm{Ass}(R^*)$. Then $x_1 \notin \cup \mathrm{Ass}(R^*)$, so that $x_1 \notin \cup \mathrm{Ass}(R) = \cap A^*(J)$, $J \sim 0$. Thus x_1 is a strong asymptotic sequence. If x_1, x_2 is not a strong asymptotic sequence, then let $P \in \cap A^*(J)$, $J \sim x_1R$ with $x_2 \in P$. Clearly $x_1 \in P$, making P regular. By Proposition 10.11 ((v) $\Rightarrow$ (ii)), $(R_P)^*$ has a depth 1 prime divisor of zero. By Corollary 10.16, there are primes $Q^* \subset P^*$ in R^* with $P^* \cap R = P$, $Q^* \in \mathrm{Ass}(R^*)$ and height $P^*/Q^* = 1$. Since $(x_1, x_2) \subseteq P \subseteq P^*$, we get $\mathrm{height}(x_1, x_2)R^* + Q^*/Q^* =$ height $P^*/Q^* = 1$, a contradiction.

CHAPTER XI: Miscellaneous

Rees' Valuations

Our first topic is due to Rees [Rs1]. The crucial primes used in Proposition 11.5 are the ones mentioned in Proposition 3.18(iii), and this material will produce a characterization of projective equivalence. We offer these facts as our excuse for reproducing this lovely mathematics.

DEFINITION. Let I be an ideal of an arbitrary ring R, and let $x \in R$. Then $V_I(x) = n < \infty$ means $x \in I^n - I^{n+1}$, while $V_I(x) = \infty$ means $x \in \cap I^n$, $n = 1,2,3,\ldots$.

PROPOSITION 11.1. Let I be an ideal in any ring R and let $x \in R$. Then $\bar{V}_I(x) = \lim_{n\to\infty} \frac{V_I(x^n)}{n}$ exists (possibly being infinite).

Proof: Let $\alpha = \limsup \frac{V_I(x^n)}{n}$. Let $\beta < \alpha$ and $\delta < 1$. We will show that for all sufficiently large n, $\frac{V_I(x^n)}{n} \geq \delta\beta$. Since δ can be arbitrarily close to 1, and β can be arbitrarily close to α, this will show that $\bar{V}_I(x)$ exists.

Since $\beta < \alpha$, there is an m with $\frac{V_I(x^m)}{m} \geq \beta$. Choose ℓ an integer with $\frac{\ell}{\ell+1} \geq \delta$, and consider $n \geq \ell m$. Suppose that $km \leq n < (k+1)m$ with k an integer. Note that $k \geq \ell$ so $\frac{k}{k+1} \geq \delta$. Since $V_I(x^m) \geq m\beta$, we clearly have $V_I(x^n) \geq V_I(x^{km}) \geq k\,V_I(x^m) \geq km\beta$. This, together with $n < (k+1)m$, gives $\frac{V_I(x^n)}{n} > \frac{km\beta}{m(k+1)} = \frac{k}{k+1}\beta \geq \delta\beta$. As this holds for all $n \geq \ell m$, we are done.

PROPOSITION 11.2. Let $k \geq 0$ be an integer if $x \in \overline{I^k}$, then $\bar{V}_I(x) \geq k$. If $\bar{V}_I(x) > k$, then $x \in \overline{I^k}$.

Proof: If $x \in \overline{I^k}$, then we have $x^{s+1} + a_1x^s + \ldots + a_{s+1} = 0$ with $a_i \in I^{ki}$. This easily shows that for $m \geq s+1$, $x^m \in I^{(m-s)k}$. Thus $\frac{V_I(x^m)}{m} \geq \frac{(m-s)}{m}k$. Letting $m \to \infty$, we see that $\bar{V}_I(x) \geq k$.

If $\bar{V}_I(x) > k$, then for large n we have $\frac{V_I(x^n)}{n} > k$ so that $x^n \in I^{nk}$. Thus

$x^n + b_1 x^{n-1} + \ldots + b_n = 0$ where $b_n = -x^n \in (I^k)^n$ and the other coefficients are 0. This shows that $x \in \overline{I^k}$.

What happens in Proposition 11.2 in case $\overline{V}_I(x) = k$? We will show that in a Noetherian ring, it follows that $x \in \overline{I^k}$. First, however, we show that in a non-Noetherian ring this need not by the case. For this, consider $K(X,Y)$, with K a field and X and Y indeterminants. Let w be a valuation on $K(X,Y)$ with values in $Z \times Z$ ordered lexicographically. Let $w(K) = (0,0)$, $w(X) = (1,0)$ and $w(Y) = (1,1)$. Let W be the valuation ring of w and let $I = YW$. Since $w(X) < w(Y)$, $X \notin I$. Since for all $n \geq 1$, $(n-1,n-1) < (n,0) < (n,n)$, we have $X^n \in I^{n-1} - I^n$. Thus $V_I(X^n) = n-1$ and so $\overline{V}_I(X) = 1$. However, I is principal and W is integrally closed, so $\overline{I} = I$, and $X \notin \overline{I}$.

LEMMA 11.3. Let R be a Krull domain and let $I = uR$ be a principal ideal. Let $p_1, \ldots, p_r$ be the height 1 primes of R containing u, and let v_i, $i = 1, \ldots, r$, be the valuation associated with the D.V.R. R_{p_i}. Let $e_i = v_i(u)$. Then for $x \in R$, $\overline{V}_I(x) = \min\{\frac{v_i(x)}{e_i} \mid i = 1, \ldots, r\}$.

Proof: Let $\alpha = \min\{\frac{v_i(x)}{e_i} \mid i = 1, \ldots, r\}$ and let $V_I(x) = k$. We claim that $\alpha - 1 < k \leq \alpha$. Since $x \in I^k = u^k R$, we have $v_i(x) \geq v_i(u^k) = ke_i$, showing $\alpha \geq k$. Now let m be an integer with $\alpha - 1 < m \leq \alpha$. Then for $i = 1, \ldots, r$, we have $v_i(u^m) = me_i \leq \alpha e_i \leq v_i(x)$. This shows that $x \in u^m R = I^m$ so that $k = V_I(x) \geq m > \alpha - 1$. Our claim is now proved. Applying it to x^n, and noting that $v_i(x^n) = nv_i(x)$, we have $n\alpha - 1 < V_I(x^n) \leq n\alpha$. Dividing by n and letting $n \to \infty$, we have $\overline{V}_I(x) = \alpha$.

LEMMA 11.4. Let I be an ideal in any domain R, and let the domain T be an integral extension of R. Let $J = IT$. For $x \in R$, $\overline{V}_I(x) = \overline{V}_J(x)$.

Proof: Since $I^n \subseteq J^n$, clearly $\overline{V}_I(x) \leq \overline{V}_J(x)$. For the reverse, we first claim that $J^n \cap R \subseteq \overline{I^n}$. For this, let $y \in J^n \cap R$, writing $y = a_1 t_1 + \ldots + a_m t_m$ with $a_i \in I^n$ and $t_i \in T$. Let $w_1, \ldots, w_k$ be module generators for $T_1 = R[t_1, \ldots, t_m]$ over R. Since $yw_j \in I^n T_1$, $yw_j = b_{j1} w_1 + \ldots + b_{jk} w_k$ with $b_{j\ell} \in I^n$. A determinant argument now proves the claim.

Consider $\beta < \bar{V}_J(x)$ with β rational. There are large integers n with $V_J(x^n) \geq n\beta$ and with $n\beta$ integral. Thus $x^n \in J^{n\beta} \cap R \subseteq \overline{I^{n\beta}}$. By Proposition 11.2, $\bar{V}_I(x^n) \geq n\beta$. Thus $\bar{V}_I(x) = \dfrac{\bar{V}_I(x^n)}{n} \geq \beta$. Letting $\beta \to \bar{V}_J(x)$, we see that $\bar{V}_I(x) \geq \bar{V}_J(x)$, completing the proof.

PROPOSITION 11.5. Let I be an ideal in a Noetherian domain R. Let $\mathcal{R} = R[t^{-1}, It]$ be the Rees ring of R with respect to I. Let $u = t^{-1}$. Let $p_1, \ldots, p_r$ be the height 1 primes of $\bar{\mathcal{R}}$ which contain u, and for $i = 1, \ldots, r$, let v_i be the valuation associated with the D.V.R. $\bar{\mathcal{R}}_{p_i}$. Let $e_i = v_i(u)$. Then for $x \in R$,
$$\bar{V}_I(x) = \min\{\dfrac{v_i(x)}{e_i} \mid i = 1, \ldots, r\}.$$

Proof: Since $u^m \mathcal{R} \cap R = I^m$, for all $m \geq 0$, clearly $V_I(x^n) = V_{u\mathcal{R}}(x^n)$ for all $n \geq 0$. Thus for $x \in R$, $\bar{V}_I(x) = \bar{V}_{u\mathcal{R}}(x)$. Now by Lemma 11.4 we see that $\bar{V}_I(x) = \bar{V}_{u\bar{\mathcal{R}}}(x)$. We now use Lemma 11.3.

COROLLARY 11.6. If I is an ideal in a Noetherian domain R and if $k \geq 0$ is an integer, then $x \in \overline{I^k}$ if and only if $\bar{V}_I(x) \geq k$.

Proof: According to Proposition 11.2 we must only show that if $\bar{V}_I(x) = k$, then $x \in \overline{I^k}$. With notation as in Proposition 11.5, we have $v_i(x) \geq e_i k = v_i(u^k)$, $i = 1, \ldots, r$. Thus $x \in u^k \bar{\mathcal{R}} \cap R = \overline{I^k}$.

We now drop the assumption that R be a domain.

PROPOSITION 11.7. Let R be a Noetherian ring with minimal primes $q_1, \ldots, q_s$. Let I be an ideal of R and let $x \in R$. For $i = 1, \ldots, s$ let $I_i = I + q_i / q_i$ and let $x_i = x + q_i$. Then $\bar{V}_I(x) = \min\{\bar{V}_{I_i}(x_i) \mid i = 1, \ldots, s\}$.

Proof: Let $\beta \leq \min\{\bar{V}_{I_i}(x_i)\}$ with β rational, and let n be a positive integer with $n\beta$ integral. Now $n\beta \leq \min\{\bar{V}_{I_i}(x_i^n)\}$. Since R/q_i is a domain, Corollary 11.6 shows that $x_i^n \in \overline{I_i^{n\beta}}$. Thus $I_i^{n\beta}$ reduces $(I_i^{n\beta}, x_i^n)$. By Lemma 3.6, $I^{n\beta}$ reduces $(I^{n\beta}, x^n)$, showing that $x^n \in \overline{I^{n\beta}}$. By Proposition 11.2 $\bar{V}_I(x^n) \geq n\beta$. Thus $\bar{V}_I(x) =$

$\frac{\overline{V}_I(x^n)}{n} \geq \beta$. This gives $\overline{V}_I(x) \geq \min\{\overline{V}_{I_i}(x_i) \mid i=1,\ldots,s\}$.

Now suppose $\gamma < \overline{V}_I(x)$ with γ rational. For infinitely many positive integers m, $V_I(x^m) > m\gamma$ and $m\gamma$ is integral. Thus $x^m \in I^{m\gamma}$ so that for $i=1,\ldots,s$, $x_i^m \in I_i^{m\gamma}$, giving $V_{I_i}(x_i^m) \geq m\gamma$. As $m \to \infty$, we have $\overline{V}_{I_i}(x_i) \geq \gamma$, which shows that $\min\{\overline{V}_{I_i}(x_i) \mid i=1,\ldots,s\} \geq \overline{V}_I(x)$.

COROLLARY 11.8. Let I be an ideal in a Noetherian ring R. For $x \in R$, and $k \geq 0$ an integer, $\overline{V}_I(x) \geq k$ if and only if $x \in \overline{I^k}$.

Proof: This follows easily from Corollary 11.6 and Lemma 3.6.

COROLLARY 11.9. Let I and J be ideals in a Noetherian ring R.

i) $\overline{I} = \overline{J}$ if and only if $\overline{V}_I(x) = \overline{V}_J(x)$ for all $x \in R$.

ii) $\overline{I^n} = \overline{J^m}$ if and only if $m\overline{V}_I(x) = n\overline{V}_J(x)$ for all $x \in R$.

Proof: i) Suppose $\overline{V}_I(x) = \overline{V}_J(x)$ for all $x \in R$. Using Corollary 11.8 we have $x \in \overline{I}$ if and only if $\overline{V}_I(x) \geq 1$ if and only if $\overline{V}_J(x) \geq 1$ if and only if $x \in \overline{J}$. Conversly, suppose that $\overline{I} = \overline{J}$. Then for all $k \geq 1$, $\overline{I^k} = \overline{\overline{I}^k} = \overline{\overline{J}^k} = \overline{J^k}$. Suppose now that for some x, $\overline{V}_I(x) > \overline{V}_J(x)$. We will derive a contradiction. For sufficiently large n, there will be an integer k such that $\overline{V}_I(x^n) = n\overline{V}_I(x) \geq k > n\overline{V}_J(x) = \overline{V}_J(x^n)$. By Corollary 11.8, $x^n \in \overline{I^k}$ but $x^n \notin \overline{J^k}$, the desired contradiction.

ii) We first claim that for any $n \geq 0$, and any $x \in R$, $n\overline{V}_{I^n}(x) = \overline{V}_I(x)$. For this, let $\beta < \overline{V}_I(x)$ with β rational. There are infinitely many m with $m\beta$ integral and with $V_I(x^{nm}) > nm\beta$. Thus $x^{nm} \in I^{nm\beta}$ so that $V_{I^n}(x^{nm}) \geq m\beta$. As $m \to \infty$, we have $\overline{V}_{I^n}(x) \geq \beta/n$. Therefore $n\overline{V}_{I^n}(x) \geq \overline{V}_I(x)$. The reverse inequality is proved similarly.

Suppose now that $m\overline{V}_I(x) = n\overline{V}_J(x)$ for all $x \in R$, with n and m integers. Then $\overline{V}_{I^n}(x) = \overline{V}_I(x)/n = \overline{V}_J(x)/m = \overline{V}_{J^m}(x)$. By (i), $\overline{I^n} = \overline{J^m}$. The converse is proved similarly.

A Question of Krull

We will use some of our knowledge of asymptotic sequences to construct an example concerning an old question. In 1937, Krull [Kr, p. 755] asked if $R \subseteq T$ is an integral extension of domains with R normal and if $P \subset Q$ are adjacent primes in T (that is, height $Q/P = 1$) must $P \cap R \subset Q \cap R$ be adjacent in R? In 1973, Kaplansky gave a negative answer [K2]. Kaplansky's example has R non-Noetherian, and he pointed out that the construction of a Noetherian example would probably require a counterexample to the chain conjecture. In 1980 such a counterexample was found (see [H2]) and shortly thereafter Ratliff used it to construct a Noetherian counterexample to Krull's question. Although the construction does not require the use of asymptotic sequences, using them gives a short path to the example.

DEFINITION. The domain R is a G.B.-ring (going between) if any pair of adjacent primes in any integral extension domain of R, contracts to a pair of adjacent primes in R.

Two lemmas will proceed our example. The first shifts attention from integral extensions of R to the polynomial ring $R[Y]$.

LEMMA 11.10. Let R be a Noetherian domain and let Y be indeterminate. Let $P \subset Q$ be primes of R with height $Q/P > 1$. In $R[Y]$, let $p \subset q$ be primes with $p \cap R = P$ and $q \cap R = Q$ and with height $q/p = 1$. If p contains a monic polynomial, then R is not a G.B. ring.

Proof: Let p contain the monic polynomial $f(Y)$. In some splitting field, let $f(Y) = (Y-u_1) \ldots (Y-u_n)$, and let $S = R[u_1, \ldots, u_n]$. Thus $R \subseteq S$ is an integral extension. In $S[Y]$, let p' be a prime lying over p. Since $f(Y) \in p'$, for some i we have $Y - u_i \in p'$. Thus $(Y-u_i)S[Y] \cap R[Y] \subseteq p$. Now $(Y-u_i)S[Y] \cap R[Y]$ is easily seen to be the kernel of the map $R[Y] \to R[u_i]$. Since this kernel is contained in p, the images under this map of $p \subset q$ are two adjacent primes of $R[u_i]$, whose contractions to R, $P \subset Q$, are not adjacent.

LEMMA 11.11. Let (R,M) be a local domain of altitude at least 3. Let a,b be elements of R with height $(a,b) = 2$. Suppose that $\overline{R[a/b]}$ contains a height 1 prime M' lying over M. Then with X an indeterminate, $R[X]$ is not a G.B.-ring.

Proof: Let $x = a/b$ and pick $y \in M'$ with y not in any other prime of $\overline{R[x]}$ which lies over $M' \cap R[x]$. Then $N = M' \cap R[x,y]$ is a height 1 prime of $R[x,y]$. Let p be the kernel of $R[X][Y] \to R[x,y]$ sending X to x and Y to y. Let q be the inverse image of N. Thus $p \subset q$ and height $q/p = 1$. Letting $P = p \cap R[X]$ and $Q = q \cap R[X]$, to invoke Lemma 11.10, we must show that height $Q/P > 1$, since y integral over $R[x]$ shows that p contains a polynomial monic in Y. Now P is the kernel of the map $R[X] \to R[x]$ and this map clearly sends Q to $N \cap R[x]$. Thus we have height $Q/P =$ height $N \cap R[x]$. As $M \subseteq N$, $MR[x] \subseteq N \cap R[x]$. By [D, Corollary 2], $MR[x]$ is a prime with height equal to $\dim R - 1 \geq 2$. Thus height $Q/P \geq 2$ and by Lemma 11.10 we are done.

EXAMPLE. Let (R,M) be a local 3-dimensional normal domain not satisfying the altitude formula. Thus R is not quasi-unmixed, so $z(M) < 3$. As R is normal, $z(M) > 1$ (Proposition 3.19). Therefore $z(M) = 2$. Using Proposition 5.6, let a,b be a maximal asymptotic sequence in R. Obviously height $(a,b) = 2$. Since $M \in \bar{A}^*(a,b)$, by Proposition 3.20, we may assume that $\overline{R[a/b]}$ contains a height 1 prime lying over M. By Lemma 11.11, $R[X]$ is a normal Noetherian domain which is not a G.B.-ring.

Remark: $R[X_1, \ldots, X_n]$ not being a G.B.-ring is thoroughly studied in [R5] and [B4].

z(P) and Conforming Pairs

PROPOSITION 11.12. Let (P,W) be a conforming pair in a Noetherian ring R, and suppose that $z(Q) \leq n+1$ for all $Q \in W$. Then $z(P) \leq n$.

Proof: Suppose that $z(P) > n-1$. Since $z(P) = \mathrm{gr}^* P_P$, let $a_1, \ldots, a_n$ be elements of P whose images in R_P form an asymptotic sequence. We will show that it is a maximal asymptotic sequence, so that $z(P) = n$. We first claim that for all but finitely many $Q \in W$, the following is True: If $p \in \cup \overrightarrow{A}^*((a_1, \ldots, a_i))$, $i = 0,1,\ldots,n$, and $p \in Q$, then $p \subseteq P$. This is easily seen since $\cup \overrightarrow{A}^*((a_1, \ldots, a_i))$ $i = 0,1,\ldots,n$, is a finite set, and (since (P,W) is a conforming pair) any $p \not\subseteq P$ is contained in at most finitely many $Q \in W$.

By deleting the finitely many exceptions in the above claim, we add the assumption that for all $Q \in W$, $p \in \cup \overrightarrow{A}^*((a_1, \ldots, a_i))$, $i = 0,1,\ldots,n$, if $p \subseteq Q$, then $p \subseteq P$. We now claim that for all $Q \in W$, the images of $a_1, \ldots, a_n$ are an asymptotic sequence in R_Q. If not, then for some $Q \in W$ and $i = 0,1,\ldots,n-1$ we have a prime $p_Q \in \overrightarrow{A}^*((a_1, \ldots, a_i)R_Q)$ with the image of a_{i+1} in p_Q. Note that $a_{i+1} \in p \in \overrightarrow{A}^*((a_1, \ldots, a_i))$ and $p \subseteq Q$. By the above assumption, $p \subseteq P$. Thus $p_P \in \overrightarrow{A}^*((a_1, \ldots, a_i)R_P)$ and the image of a_{i+1} is in p_P. This contradicts that the images of $a_1, \ldots, a_n$ form an asymptotic sequence in R_P, and proves our second claim.

Finally, we claim that $P \in \overrightarrow{A}^*((a_1, \ldots, a_n))$. If not, pick $a_{n+1} \in P$ with a_{n+1} not in any prime p satisfying $p \in \overrightarrow{A}^*(a_1, \ldots, a_n)$ and $p \subset P$. Our assumption on W easily shows that the images of $a_1, \ldots, a_{n+1}$ form an asymptotic sequence in R_Q for all $Q \in W$. Since $gr^* Q_Q = z(Q) \leq n+1$, this must be a maximal asymptotic sequence, showing that $Q_Q \in \overrightarrow{A}^*((a_1, \ldots, a_{n+1})R_Q)$, and $Q \in \overrightarrow{A}^*(a_1, \ldots, a_{n+1})$, for all $Q \in W$. As W is infinite, this is impossible, proving our final claim. We now have $P \in \overrightarrow{A}^*(a_1, \ldots, a_n)$ so that the images of $a_1, \ldots, a_n$ are a maximal asymptotic sequence in R_P. Thus $z(P) = gr^*(P_P) = n$.

The above argument, due to Heitmann, similarly shows that if grade $Q_Q \leq n+1$ for all $Q \in W$, then grade $P_P \leq n$.

COROLLARY 11.13. Let P be a prime in a Noetherian ring R such that R_P is quasi-unmixed. Let $W = \{Q \in \text{spec } R \mid P \subset Q, \text{ height } Q/P = 1\}$. Then for all but finitely many $Q \in W$, R_Q is quasi-unmixed.

Proof: If W is finite there is no problem. Suppose W is infinite. Then (P,W) is a conforming pair. Let height $P = n$. As R_P is quasi-unmixed, $z(P) = n$. Let $W' = \{Q \in W \mid z(Q) \leq n\}$. If W' is infinite, then (P,W') is a conforming pair, and Proposition 11.12 shows that $z(P) \leq n-1$, a contradiction. Thus W' is finite. Now let $W'' = \{Q \in W \mid \text{height } Q > n+1\}$. By Lemma 9.11, W'' is finite. Since we always have $z(Q) \leq \text{height } Q$, for any $Q \in W - (W' \cup W'')$ we have $n+1 \leq z(Q) \leq \text{height } Q \leq n+1$. Thus $z(Q) = \text{height } Q$, and R_Q is quasi-unmixed.

The recently constructed counterexample to the chain conjecture gives a normal Noetherian domain with primes $P \subset Q$ such that height $P = 2$, height $Q/P = 1$ and little height $Q = 2$. As another corollary of the above, we show that for a fixed height 2 prime P in a normal Noetherian domain, there can be at most finitely many such Q. (If R is not normal, there can be infinitely many such Q.)

COROLLARY 11.14. Let $P \neq 0$ be a prime in the Noetherian domain R. Let $W = \{Q \in \text{spec } R \mid P \subset Q, \text{ height } Q/P = 1, \text{ and little height } Q = 2\}$. If W is infinite, the $\overline{R}$ contains a height 1 prime lying over P. If R is normal, height $P = 1$.

Proof: Since $z(Q) \leq$ little height Q (see the Appendix), we have $z(Q) \leq 2$ for all $Q \in W$. If W is infinite, then clearly (P,W) is a conforming pair. By Proposition 11.12, $z(P) = 1$ (since $P \neq 0$ implies $z(P) \neq 0$). Thus $(R_P)^*$ contains a depth 1 minimal prime. By Proposition 3.19, $\overline{R}$ contains a height 1 prime lying over P. The last sentence of the Corollary is obvious.

Question: Suppose for primes $P \subseteq Q$ in a Noetherian ring, we define the relation z-catenicity to mean $z(Q) = z(P) + z(Q/P)$. Is z-catenicity a conforming relation?

Prenormality and $\cup A^*(J)$, $J \sim I$

Recalling that an ideal I is normal if $\overline{I^n} = I^n$ for all $n \geq 1$, we make the following definition.

DEFINITION. The ideal I is prenormal if some power of I is normal.

In $K[X^3,X^4,X^5]$, K a field, $I = (X^3,X^4)$ is prenormal but not normal since $X^5 \in \overline{I} - I$, while $\overline{I^n} = I^n$ for all $n \geq 2$.

PROPOSITION 11.15. Let I be a regular ideal in a Noetherian ring. The following are equivalent.

i) I is prenormal.

ii) $I^n = \overline{I^n}$ for infinitely many n.

iii) $I^n = \overline{I^n}$ for all large n.

iv) $\widetilde{I^n} = \overline{I^n}$ for all large n.

v) $\widetilde{I^n} = \overline{I^n}$ for all n.

Proof: (i) $\Rightarrow$ (ii), and (iii) $\Rightarrow$ (i) are obvious. For (ii) $\Rightarrow$ (iii) we use Lemma 8.1 and have $(\overline{I^{n+\ell}} : I^\ell) = \overline{I^n}$ for all large n and any ℓ. Since $\overline{I^n} I^\ell \subseteq \overline{I^{n+\ell}}$, and since by (ii) we may choose ℓ such that $\overline{I^{n+\ell}} = I^{n+\ell}$, we see that $\overline{I^n} \subseteq (I^{n+\ell} : I^\ell) = I^n$, proving (iii). For (iii) $\Rightarrow$ (v), for any n, select m large enough that $I^{nm} = \overline{I^{nm}}$. Now $(I^n)^m \subseteq (\overline{I^n})^m \subseteq \overline{I^{nm}} = I^{nm}$, so that $(I^n)^m = (\overline{I^n})^m$. By Proposition 8.2(iv), $\overline{I^n} \subseteq \widetilde{I^n}$. Since the other inclusion is automatic, we have (v). Obviously (v) $\Rightarrow$ (iv), and (iv) $\Rightarrow$ (iii) follows from Proposition 8.2(ii).

PROPOSITION 11.16. Let I be a regular ideal in a Noetherian ring R. Then $\mathrm{Ass}(\overline{I^n}/I^n)$ stabilizes for large n.

Proof: For large n, $(I^{n+1} : I) = I^n$. Suppose that $P \in \mathrm{Ass}(\overline{I^n}/I^n)$ and write $P = (I^n : c)$ with $c \in \overline{I^n}$. Clearly $P \subseteq (I^{n+1} : cI)$, and if $x \in (I^{n+1} : cI)$ then $xc \in (I^{n+1} : I) = I^n$, so that $x \in (I^n : c) = P$. This shows $P = (I^{n+1} : cI)$. As $cI \subseteq \overline{I^n} I \subseteq \overline{I^{n+1}}$, we have $P \subseteq \mathrm{Ass}(\overline{I^{n+1}}/I^{n+1})$. Therefore, for large n, $\mathrm{Ass}(\overline{I^n}/I^n)$ is an increasing sequence. Also, for large n, $\mathrm{Ass}(\overline{I^n}/I^n) \subseteq \mathrm{Ass}(R/I^n) \subseteq A^*(I)$. Since $A^*(I)$ is finite, the result follows.

DEFINITION. For I a regular ideal in a Noetherian ring, let $C^*(I)$ denote the limit of the sequence $\mathrm{Ass}(\overline{I^n}/I^n)$, $n = 1, 2, \ldots$

We now see that prenormality is a local condition.

PROPOSITION 11.17. Let I be a regular ideal in a Noetherian ring. Then I is prenormal if and only if I_P is prenormal for all $P \in C^*(I)$.

Proof: One direction is trivial. Thus suppose $C^*(I) = \{P_1, \ldots, P_m\}$ and that n is large enough that $\mathrm{Ass}(\overline{I^k}/I^k) = C^*(I)$ and $I^k_{P_i} = \overline{I^k_{P_i}}$ for all i and for all $k \geq n$. We claim that $\overline{I^k} = I^k$ for all $k \geq n$. If not, pick $x \in \overline{I^k} - I^k$. Then $(I^k : x)$ is a proper ideal. By Lemma 1.2, for some $d \in R$, $(I^k : xd)$ is a prime P. As $xd \in \overline{I^k}$, $P \in \mathrm{Ass}(\overline{I^k}/I^k) = C^*(I)$. Since $x \in \overline{I_P^k} = I_P^k$, there is an $s \in R - P$ with $sx \in I^k$, contradicting that $(I^k : x) \subseteq P$.

Since $C^*(I)$ is finite, the next result shows that $\{Q \in \mathrm{spec}\, R \mid I_Q$ is prenormal$\}$ is open in the spec topology.

PROPOSITION 11.18. Let I be a regular ideal in a Noetherian ring R. Then for $Q \in \mathrm{spec}\, R$, I_Q is not prenormal if and only if there is a $P \in C^*(I)$ with $P \subseteq Q$.

Proof: I_Q is prenormal if and only if $I_Q^n = \overline{I_Q^n}$ for all big n if and only if $\mathrm{Ass}(\overline{I_Q^n}/I_Q^n) = \emptyset$ for all big n if and only if $C^*(I_Q) = \emptyset$. If $C^*(I_Q) \neq \emptyset$, then there is a $P_Q \in C^*(I_Q)$. We easily see that $P \in C^*(I)$ and $P \subseteq Q$. Conversly, if $C^*(I_Q) = \emptyset$ then no such P can exist.

PROPOSITION 11.19. Let I be a regular ideal in a Noetherian ring. Then $A^*(I) = \overrightarrow{A}^*(I) \cup C^*(I)$.

Proof: One containment is obvious. Thus let $P \in A^*(I) - \overrightarrow{A}^*(I)$. Therefore $P_P \in A^*(I_P) - \overrightarrow{A}^*(I_P)$, so that for big n we have $P_P = (I_P^n : c_n)$ with $c_n \in R_P$. If $c_n \notin \overline{I_P^n}$, then $P_P \in \overrightarrow{A}^*(I_P)$ contradicting that $P \notin \overrightarrow{A}^*(I)$. Thus $c_n \in \overline{I_P^n}$, and so $P_P \in C^*(I_P)$, showing that $P \in C^*(I)$ as desired.

In Chapter 10 we looked at $\cap A^*(J)$, $J \sim I$. Now we glance at $\cup A^*(J)$, $J \sim I$, and see that in many circumstances this union is big.

LEMMA 11.20. Let I be a regular ideal in a Noetherian ring R. If P is a prime containing I, and if I_P is not prenormal, then $P \in \cup A^*(J)$, $J \sim I$.

Proof: We have $I_P^n \neq \overline{I_P^n}$ for all big n. By Proposition 8.2(ii), $I_P^n = \widetilde{I_P^n}$ for all big n. Thus for big n, $\overline{I_P^n} \not\subseteq \widetilde{I_P^n}$, and Lemma 8.3 gives $P_P \in A^*(I_P^n + P_P\overline{I_P^n})$. Thus $P \in A^*(I^n + P\overline{I^n})$. As $I^n + P\overline{I^n} \sim I$, we are done.

PROPOSITION 11.21. Let $I=(a_1,\ldots,a_n)$ be a regular ideal in a Noetherian ring with $n>1$. Suppose that P is a prime containing I, and that in R_P, $a_1,\ldots,a_n$ are analytically independent. Then $P\in \cup A^*(J)$, $J\sim I$.

Proof: Let $\hat{I}=(a_1^2,\ldots,a_n^2)$. In R_P, $a_1^2,\ldots,a_n^2$ are also analytically independent. Now $\hat{I}\subseteq I^2\subseteq \bar{\hat{I}}$. For any $k\geq 1$, $\hat{I}_P^k$ is minimally generated by all monomials of degree k in $a_1^2,\ldots,a_n^2$, while I_P^{2k} is minimally generated by all monomials of degree $2k$ in $a_1,\ldots,a_n$. Thus $v(\hat{I}_P^k)\neq v(I_P^{2k})$, so that $\hat{I}_P^k\neq (I_P^2)^k$. As $\hat{I}_P^k\subseteq (I_P^2)^k\subseteq \overline{(\hat{I}_P^k)}$, we have $\hat{I}_P^k\neq \overline{(\hat{I}_P^k)}$, for all $k\geq 1$, showing that $\hat{I}_P$ is not pre-normal. By Lemma 11.20, $P\in\cup A^*(J)$, $J\sim\hat{I}$. However, $\hat{I}\subseteq I^2\subseteq\bar{\hat{I}}$ shows that $\hat{I}\sim I$, so we are done.

COROLLARY 11.22. Let I be a regular ideal in the principal class of a Noetherian ring. If height $I=n>1$, then for any prime P containing I, $P\in\cup A^*(J)$, $J\sim I$.

Proof: Let $I=(a_1,\ldots,a_n)$. If P contains I, then height $I_P=n$ shows that $a_1,\ldots,a_n$ are analytically independent in R_P, and we use Proposition 11.21.

COROLLARY 11.23. Let (R,M) be a local domain with R/M infinite. Let I be a regular ideal with $\ell(I)=n>1$. Then $M\in\cup A^*(J)$, $J\sim I$.

Proof: Let $\hat{I}=(a_1,\ldots,a_n)$ be a minimal reduction of I. As $a_1,\ldots,a_n$ are analytically independent, Proposition 11.21 gives $M\in\cup A^*(J)$, $J\sim\hat{I}$. However $\hat{I}\sim I$.

Remark: We do not know if this works for a prime other than the maximal ideal of a local ring, the problem being that $\hat{I}_P$ a reduction of I_P, does not imply that $\hat{I}$ reduces I. The next section is related.

Local Projective Equivalence

Throughout this section, R will always be Noetherian, and all ideals will be assumed to have height greater than zero.

LEMMA 11.24. Suppose $I\sim J$, with $\overline{I^n}=\overline{J^m}$, n and m positive integers. Then if k and ℓ are positive integers, $\overline{I^k}=\overline{J^\ell}$ if and only if $n/m=k/\ell$.

Proof: Assume that $\overline{I^k} = \overline{J^\ell}$. By Corollary 11.9, $(n/m)\overline{V}_J(x) = \overline{V}_I(x) = (k/\ell)\overline{V}_J(x)$ for all $x \in R$. In order to conclude that $n/m = k/\ell$, we need only exhibit an x with $0 < V_J(x) < \infty$. As height $J > 0$, pick $x \in J$ but in no minimal prime. By Corollary 11.8, $\overline{V}_J(x) \geq 1 > 0$. By Lemma 3.11, $x \notin \cap \overline{I^j}$, $j = 1,2,3,\ldots$, so that Corollary 11.8 also gives $\overline{V}_J(x) < \infty$. (Note: finding this x is the purpose of our restriction that height J be at least 1. In fact such an x exists if for some maximal $M \geq J$, J_M is not in the nilradical of R_M. However, this last condition does not localize.)

DEFINITION. If $I \sim J$, let $\gamma(I,J) = n/m$ with $\overline{I^n} = \overline{J^m}$.

LEMMA 11.25. Let $I \sim J$ and let P be a prime containing I. Then $I_P \sim J_P$ and $\gamma(I_P, J_P) = \gamma(I,J)$.

Proof: If $\overline{I^n} = \overline{J^m}$, then $\overline{I_P^n} = \overline{J_P^m}$.

EXAMPLE. Let M and N be distinct maximal ideals. Let $I = M^2N$ and $J = MN^2$. Then $I_M = M_M^2 = J_M^2$ and $I_N^2 = N_N^2 = J_N$. Thus $I_M \sim J_M$ and $I_N \sim J_N$. Yet $I \not\sim J$ since if $I \sim J$, then Lemma 11.25 would give $\gamma(I_M, J_M) = \gamma(I,J) = \gamma(I_N, J_N)$. However, obviously $\gamma(I_M, J_M) = 1/2 \neq 2 = \gamma(I_N, J_N)$.

The preceding example illustrates the main problem in going from local to global projective equivalence, as we now see.

THEOREM 11.26. The following are equivalent

(i) $I \sim J$

(ii) $I_M \sim J_M$ for all maximal ideals M containing $I \cap J$, and $\{\gamma(I_M, J_M) \mid I \cap J \subseteq M \text{ maximal}\}$ has size 1.

(iii) $I_P \sim J_P$ for all $P \in \overline{A}^*(I) \cup \overline{A}^*(J)$, and $\{\gamma(I_Q, J_Q) \mid Q \text{ is minimal over } I\}$ has size 1.

Proof: (i) $\Rightarrow$ (ii) $\Rightarrow$ (iii). Both follow from Lemma 11.25. Suppose (iii) holds, with $\{\gamma(I_Q, J_Q) \mid Q \text{ is minimal over } I\} = \{n/m\}$. We will show that $\overline{I^n} = \overline{J^m}$, proving (i). If false, suppose $x \in \overline{I^n} - \overline{J^m}$ (the other possibility being similar). Then

$\overline{(J^m : x)}$ is a proper ideal, which by Lemma 1.2 can be expanded to a prime $P = (\overline{J^m} : xy)$ for some $y \in R$. Now $P \in \bar{A}(J,m) \subseteq \bar{A}^*(J)$ and so by (iii), $I_P \sim J_P$. This shows that $\operatorname{rad} I_P = \operatorname{rad} J_P$, so there is a prime $Q \subseteq P$ with Q minimal over both I and J. We already have $\gamma(I_Q, J_Q) = n/m$, and so by Lemma 11.25, $\gamma(I_P, J_P) = n/m$. Now Lemma 11.24 gives $\overline{I_P^n} = \overline{J_P^m}$. As $x \in \overline{I^n}$, for some $s \in R - P$, $sx \in \overline{J^m}$. Thus $s \in (\overline{J^m} : x) \subseteq (\overline{J^m} : xy) = P$, a contradiction.

$\bar{A}(I,n) = \bar{B}(I,n)$

In Chapter 1 we considered the two sequences $A(I,n) = \operatorname{Ass}(R/I^n)$ and $B(I,n) = \operatorname{Ass}(I^{n-1}/I^n)$, $n = 1,2,3,\ldots$. In Chapter 3, we looked at $\bar{A}(I,n) = \operatorname{Ass}(R/\overline{I^n})$, $n = 1,2,\ldots$. Finally, we consider $\bar{B}(I,n) = \operatorname{Ass}(\overline{I^{n-1}}/\overline{I^n})$, $n = 1,2,\ldots$. We show that it is identical to $\bar{A}(I,n)$. The key is a lemma due to Katz, which will also give an easy proof that $\bar{A}(I,n)$ is increasing.

LEMMA 11.27. Let I be an ideal in a Noetherian ring R. Then for all $n \geq m \geq 1$ $(\overline{I^n} : \overline{I^m}) = (\overline{I^n} : I^m) = \overline{I^{n-m}}$.

Proof: Since $\overline{I^{n-m}} \subseteq (\overline{I^n} : \overline{I^m}) \subseteq (\overline{I^n} : I^m)$, it is enough to show $(\overline{I^n} : I^m) \subseteq \overline{I^{n-m}}$. First we assume that R is a domain, and let V be any valuation overring of R. If $x \in (\overline{I^n} : I^m)$ then $xI^mV \subseteq \overline{I^n}V = I^nV$. Since IV is principal, $x \in I^{n-m}V$. As this holds for all such V, and as $\overline{I^{n-m}} = (\cap I^{n-m}V) \cap R$, we have $x \in \overline{I^{n-m}}$. Now for an arbitrary Noetherian ring R, we let q be any minimal prime ideal, and let primes denote modulo q. Then $(\overline{I^n} : I^m)' \subseteq (\overline{I'^n} : I'^m) \subseteq \overline{I'^{n-m}}$. As this holds for any minimal prime q, Lemma 3.6 shows that $(\overline{I^n} : I^m) \subseteq \overline{I^{n-m}}$.

COROLLARY 11.28. For an ideal I in a Noetherian ring, $\bar{A}(I,n) = \bar{B}(I,n)$, $n = 1,2,$ and this sequence is increasing.

Proof: Clearly $\bar{B}(I,n) \subseteq \bar{A}(I,n)$. Let $P \in \bar{A}(I,n)$ and write $P = (\overline{I^n} : c)$, $c \in R$. Since $\bar{I} \subseteq P$, $\bar{I}c \subseteq \overline{I^n}$ showing that $c \in (\overline{I^n} : \bar{I}) = \overline{I^{n-1}}$. Thus $P \in \bar{B}(I,n)$, and $\bar{A}(I,n) = \bar{B}(I,n)$. Furthermore, clearly $P \subseteq (\overline{I^{n+1}} : c\bar{I})$ while if $x \in (\overline{I^{n+1}} : c\bar{I})$ then $xc \in (\overline{I^{n+1}} : \bar{I}) = \overline{I^n}$ so that $x \in (\overline{I^n} : c) = P$. This shows that $P = (\overline{I^{n+1}} : c\bar{I})$, so that $P \in \bar{A}(I,n+1)$, showing our sequence is increasing.

An Example

We wish to give an example to show that in Proposition 10.11 ((ii) $\Leftrightarrow$ (iv))
and in Proposition 3.19 ((i) $\Leftrightarrow$ (ii)) it is not enough to only consider prime ideals.
Thus we construct an n-dimensional local domain (R,M) such that $M \in \bar{A}^*(P)$ for
all primes $P \neq 0$ (in fact $M \in \bar{A}^*(I)$ for any ideal I containing a nonzero prime)
and $M \in \mathrm{Ass}(R/P^m)$ for all $m \geq 2$, but such that R^* has no depth 1 prime divisor of zero. First, an easy lemma.

LEMMA 11.29. Let I be an ideal in a Noetherian ring R and let $x \in I$ be a regular element. Suppose that $\bar{R} \cap R_x$ is a finite R-module. Then $T(I) \subseteq R^{[1]}$ if and
only if $T(I) \subseteq R^{\langle 1 \rangle}$.

Proof: Since $R^{\langle 1 \rangle} \subseteq R^{[1]}$, one direction is obvious. Thus suppose that $T(I) \subseteq R^{[1]}$.
Since $R^{[1]} \subseteq \bar{R}$ and $T(I) \subseteq R_x$, we have $T(I) \subseteq \bar{R} \cap R_x$, so that $T(I)$ is a
finite R-module and we use Proposition 10.16.

EXAMPLE. Let $n \geq 2$. By [HI] there is a normal Noetherian domain T satisfying
the Altitude Formula with exactly two maximal ideals N_1 and N_2, both of height
n, such that 0 is the only prime contained in $N_1 \cap N_2$. By a standard gluing
process (See [DL]) there is a local domain (R,M) satisfying the Altitude Formula
with $\bar{R} = T$, $MT \subseteq R$ (so that $\bar{R} = T$ is a finite R-module), $M = N_1 \cap N_2$, and for any
prime $P \neq M$ of R, exactly one prime of T lies over P.

For $P \in \mathrm{spec}\, R - \{0,M\}$ let Q be the unique prime of T lying over P. Since
$Q \not\subseteq N_1 \cap N_2$, without loss we may assume that $Q \subseteq N_1$ and $Q \not\subseteq N_2$. We claim that N_2
is minimal over PT. Suppose that $PT \subseteq q \subseteq N_2$ with q prime, and let $p = q \cap R$.
Clearly $P \subseteq p$, and since $Q \cap R = P$, by going up we can find a prime q' of T
with $Q \subseteq q'$ and $q' \cap R = p$. Now $q' \not\subseteq N_2$ since $Q \not\subseteq N_2$. Since $q \subseteq N_2$, $q \neq q'$
and these are two distinct primes lying over p. By the nature of $R \subseteq T$, we must
have $p = M$, so that $q' = N_2$. This shows that N_2 is minimal over PT as desired.
Therefore, by Proposition 3.5, $M \in \bar{A}^*(P)$. (Note: The same argument works for any
ideal I containing P.)

We now wish to show that $M \in \mathrm{Ass}(R/P^m)$ for $m \geq 2$. For this, choose $b_1 \in N_1 - (N_1^2 \cup N_2)$ and $b_2 \in N_2 - (N_2^2 \cup N_1)$ so that $b = b_1 b_2 \in (N_1 \cap N_2) - (N_1^2 \cup N_2^2)$. Since we already have N_2 minimal over PT, for any $m \geq 2$ we have an $s \in T - N_2$ and a $k \geq 1$ with $sN_2^k \subseteq P^m T$. Using that $b \in N_1 \cap N_2 = M$ and $MT \subseteq R$, we see that $bsM^k \subseteq bsN_2^k \subseteq bP^m T \subseteq P^m$. We claim that $bs \notin P^m$ (note $bs \in R$ since $b \in M$). If $bs \in P^m$ then $bs \in N_2^m$, and since $s \in T - N_2$, we get $b \in N_2^m$, contradicting that $b \notin N_2^2$. Thus $bsM^k \subseteq P^m$ but $bs \in R - P^m$, showing that $M \in \mathrm{Ass}(R/P^m)$, $m \geq 2$.

Finally we want that R^* does not contain a depth 1 prime divisor of zero. If it did, then by Proposition 10.11 and 10.21 we would have $T(M) \not\subseteq R^{\langle 1 \rangle}$. As $\overline{R}$ is a finite R-module, Lemma 11.29 shows that $T(M) \not\subseteq R^{[1]}$. Thus by Proposition 10.3, R^* contains a depth 1 minimal prime. As R satisfies the altitude formula, height $M = 1$. This contradicts that height $M = n \geq 2$.

Strong Asymptotic Sequences

The last paragraph of Chapter X discusses the possibility of developing a concept of strong asymptotic sequences which would stand in relation to prime divisors of zero, as asymptotic sequences stand to minimal primes. This section will discuss such a concept. As it was developed too late for major treatment in the text, this postscript will point out the path, a full exposition appearing elsewhere. I do not know if the definition of strong asymptotic sequence given here coincides with that given in Chapter X (it does for sequences of length 1 or 2).

Notation: Let $I \subseteq J$ be ideals in a Noetherian ring. $(I : J) \subseteq (I : J^2) \subseteq (I : J^2) \subseteq \ldots$ eventually stabilizes. We denote that stable ideal as $I : \langle J \rangle$. (Thanks to P. Schenzel for the nice notation, and for illustrating the significance of this ideal.)

LEMMA 11.30. Let I be an ideal in a local ring (R,M). The following are equivalent.

i) There is a $Q \in \mathrm{Ass}\, R$ with M minimal over $I + Q$.

ii) $\bigcap_{m=1}^{\infty} I^m : \langle M \rangle \neq 0$.

Further, if R is complete these are equivalent to

iii) There is a $k > 0$ such that for all $m > 0$, $I^m : \langle M \rangle \not\subseteq M^k$.

Proof: i) $\Rightarrow$ iii). If $\operatorname{Rad} I = M$, then $I^m : \langle M \rangle = R$ for all $m > 0$ and it is trivial. Thus assume $\operatorname{Rad} I \neq M$. Then $I^m \subseteq I^m : \langle M \rangle \neq R$, and so by (i), M is minimal over $I^m : \langle M \rangle + Q$. Now let $q_1 \cap \ldots \cap q_n = 0$ be a primary decomposition of 0 with $\operatorname{Rad} q_1 = Q$. Select $0 \neq x \in (q_2 \cap \ldots \cap q_n) - q_1$. Let $I^m : \langle M \rangle \subseteq P \neq M$, P prime. Then $Q \not\subseteq P$ and so we may pick $y \in q_1 - P$. As $yx = 0$, x is in every P-primary ideal. Since M is clearly not a prime divisor of $I^m : \langle M \rangle$, primary decomposition gives $x \in I^m : \langle M \rangle$. Thus (ii) holds.

In case R is complete, the equivalence of (iii) is easy using [N, 30.1] and the Krull intersection theorem.

PROPOSITION 11.31. Let $I \subseteq P$ be ideals, P prime, in a Noetherian ring R. The following are equivalent

i) There is a $q^* \in \operatorname{Ass}(R_P)^*$ with $(P_P)^*$ minimal over $I(R_P)^* + q^*$.

ii) There is an integer $k > 0$ such that $P \in \operatorname{Ass} R/J$ for every ideal J satisfying $\operatorname{Rad} J = \operatorname{Rad} I$ and $J \subseteq P^{(k)}$.

iii) There is an integer $k > 0$ such that for all $m > 0$, $I^m : \langle P \rangle \not\subseteq P^{(k)}$.

Proof: (i) $\Leftrightarrow$ (iii). Note that $I^m : \langle P \rangle \not\subseteq P^{(k)}$, $I_P^m : \langle P_P \rangle \not\subseteq P_P^k$, and $I^m(R_P)^* : \langle (P_P)^* \rangle \not\subseteq (P_P)^{*k}$ are all equivalent. Thus the equivalence of (i) and (iii) follows from Lemma 11.30.

(i) $\Rightarrow$ (ii). Choose k to equal the n of Lemma 1.13 applied to $q^* \subseteq (P_P)^*$. If $\operatorname{Rad} J = \operatorname{Rad} I$ then by (i), $(P_P)^*$ is minimal over $J(R_P)^* + q^*$. If also $J \subseteq P^{(k)}$, then $J(R_P)^* \subseteq (P_P)^{*k}$. Thus $(P_P)^*$ is a prime divisor of $J(R_P)^*$, and so $P \in \operatorname{Ass} R/J$.

(ii) $\Rightarrow$ (iii). If P is not minimal over I, then $\operatorname{Rad} I = \operatorname{Rad}(I^m : \langle P \rangle)$ and since $P \notin \operatorname{Ass} R/(I^m : \langle P \rangle)$, (iii) follows easily from (ii). If P is minimal over I, then (iii) holds with $k = 1$, using primary decomposition.

DEFINITION. For I an ideal in a Noetherian ring R, let $A_*(I) = \{P \in \operatorname{Spec} R \mid I \subseteq P$ and there is a $k > 0$ such that $P \in \operatorname{Ass} R/J$ for all ideals J satisfying $\operatorname{Rad} J = \operatorname{Rad} I$ and $J \subseteq P^{(k)}\}$.

Remarks. (a) $A_*(0) = \operatorname{Ass} R$.

(b) $A_*(I) \subseteq A^*(I)$ (and so is finite).

(c) If S is a multiplicatively closed set in R and $P \cap S = \emptyset$ then $P \in A_*(I)$ if and only if $P_S \in A_*(I_S)$. (Since $I^m : \langle P \rangle \nsubseteq P^{(k)}$ if and only if $I_S^m : \langle P_S \rangle \nsubseteq (P_S)^{(k)}$).

(d) $P \in A_*(I)$ if and only if there is a $Q \in \operatorname{Ass} R$ with $Q \subseteq P$ and $P/Q \in A_*(I+Q/Q)$ (if $P \in A_*(I)$, then there is a $q^* \in \operatorname{Ass}(R_P)^*$ with $(P_P)^*$ minimal over $I(R_P)^* + q^*$. Take $Q \in \operatorname{Ass} R$ such that $q^* \cap R_P = Q_P$).

LEMMA 11.32. Let $R \subseteq T$ be a faithfully flat extension of Noetherian rings. Let I be an ideal of R. If $P' \in A_*(IT)$, then $P' \cap R \in A_*(I)$.

Proof: Let k be as in the definition of $A_*(IT)$ applied to P'. If $\operatorname{Rad} J = \operatorname{Rad} I$ and $J \subseteq (P' \cap R)^{(k)}$, then $\operatorname{Rad} JT = \operatorname{Rad} IT$ and $JT \subseteq P'^{(k)}$ so that $P' \in \operatorname{Ass} T/JT$. Thus $P' \cap R \in \operatorname{Ass} R/J$.

The converse of Lemma 11.32 is true, but requires the following clever result.

PROPOSITION 11.33. (Schenzel) Let $I \subseteq J$ be ideals in a Noetherian ring R. Suppose for all $P \in A^*(J)$, and for all $q^* \in \operatorname{Ass}(R_P)^*$, we have $\operatorname{height} I(R_P)^* + q^*/q^* \le \operatorname{height} J(R_P)^* + q^*/q^*$. Then for all $k > 0$ there is an $m > 0$ with $I^m : \langle J \rangle \nsubseteq J^k$.

Proof: Suppose for some $k > 0$ we have $I^m : \langle J \rangle \nsubseteq J^k$ for all $m > 0$. Increasing k does no harm, and so we easily find a $P \in A^*(J)$ with $I_P^m : \langle J_P \rangle \nsubseteq J_P^k$. Therefore $I^m(R_P)^* : \langle J(R_P)^* \rangle \nsubseteq J^k(R_P)^*$. We may assume that P is minimal in $A^*(J)$ with the property that this non-containment is true for some k and all m. We claim for some $q^* \in \operatorname{Ass}(R_P)^*$, $\operatorname{height} I(R_P)^* + q^*/q^* = \operatorname{height} J(R_P)^* + q^*/q^*$. To ease notation, we will assume $R = (R_p)^*$. That is, we will assume (R,M) is a complete local ring, that $M \in A^*(J)$, that $I^m : \langle J \rangle \not\subseteq J^k$ for all $m > 0$, and that if $p \in A^*(J) - \{M\}$ then for all k there is an m with $I_p^m : \langle J_p \rangle \subseteq J_p^k$. Now let $E_m = I^m : \langle J \rangle$ and note that $E_1 \supseteq E_2 \supseteq E_3 \supseteq \dots$. Fix a k large enough that Ass R/J^k has stabilized to $A^*(J)$. Since $A^*(J) - \{M\}$ is finite, our assumption on R shows that for sufficiently large m, the module $E_m + J^k/J^k$ localized at $p \in A^*(J) - \{M\}$ is 0. Thus for large m, the only possible associated prime of $E_m + J^k/J^k$ is M. The annihilator of any element of our module therefore has radical equal to M. Since our module is finitely generated, its annihilator contains a power of M. Thus for large m, $E_m + J^k/J^k$ has finite length, and as our modules decrease as m increases, we find that for a fixed large k the ideals $E_m + J^k$ stabilize for large m. Now R is complete in the M-adic topology, hence also in the J-adic topology. The argument used to prove [N, 30.1] can be used to show that if $\bigcap_{m=1}^{\infty} E_m = 0$ then for all $k > 0$ there is an m with $E_m \subseteq J^k$. However we are assuming this containment fails for some k. Thus $\bigcap_{m=1}^{\infty} E_m \neq 0$. Let $x \neq 0$ be in that intersection. A well known corollary to the Artin-Rees Lemma shows that for large ℓ and $m \geq \ell$, $I^m : x \subseteq (0 : x) + I^{m-\ell}$. The choice of x shows that for each m there is an n with $J^n \subseteq I^m : x \subseteq (0 : x) + I^{m-\ell}$. We may expand $(0 : x)$ to a prime $q^* \in$ Ass R. Since $I \subseteq J$ and $J^n \subseteq q^* + I^{m-\ell}$ clearly Rad $I + q^*/q^* =$ Rad $J + q^*/q^*$. Thus height $I + q^*/q^* =$ height $J + q^*/q^*$. As $M \in A^*(J)$, our initial hypothesis is contradicted.

LEMMA 11.34. Let $I \subseteq J$ be ideals in a Noetherian ring R. Then there is a $k > 0$ such that for all $m > 0$, $I^m : \langle J \rangle \not\subseteq J^k$ if and only if there is a $P \in A_*(I)$ with $J \subseteq P$.

Proof: Suppose such P exists. By Proposition 11.31, there is a $k > 0$ such that for all $m > 0$, $I^m : \langle P \rangle \not\subseteq P^{(k)}$. As $I^m : \langle P \rangle \subseteq I^m : \langle J \rangle$, and $J^k \subseteq P^{(k)}$, we have $I^m : \langle J \rangle \not\subseteq J^k$. Conversely, suppose $I^m : \langle J \rangle \not\subseteq J^k$ for all $m > 0$. By Proposition 11.33, there is a prime p containing I such that for some $q^* \in \mathrm{Ass}(R_p)^*$, we have height $I(R_p)^* + q^*/q^* =$ height $J(R_p)^* + q^*/q^*$. Let $p^* \in \mathrm{Spec}(R_p)^*$ with p^*/q^* minimal over, and having the same height as

$J(R_p)^* + q^*/q^*$. As $I \subseteq J$, we also have p^* minimal over $I(R_p)^* + q^*$. Thus $p^* \in A_*(I(R_p)^*)$ by Lemma 1.13, and so by Lemma 11.32, $p^* \cap R_p \in A_*(IR_p)$. If $p^* \cap R_p = P_p$, then $I \subseteq P$ and $P \in A_*(I)$ as desired.

PROPOSITION 11.35. Let $R \subseteq T$ be a faithfully flat extension of Noetherian rings, and let I be an ideal of R. Then $P \in A_*(I)$ if and only if there is a $P' \in A_*(IT)$ with $P' \cap R = P$.

Proof: Lemma 11.32 gives one direction. Thus suppose $P \in A_*(I)$. Let $S = R - P$, so that $P_S \in A_*(I_S)$. By Proposition 11.31 there is an integer k such that for all $m > 0$, $I_S^m : \langle P_S \rangle \not\subseteq P_S^k$. Now T_S is a faithfully flat extension of R_S, and so $I_S^m T_S : \langle P_S T_S \rangle \not\subseteq P_S^k T_S$. By Lemma 11.34, there is a $P' \in \operatorname{Spec} T$ such that $P'_S \in A_*(I_S T_S)$ and $P_S T_S \subseteq P'_S$. Of course $P' \in A_*(IT)$, and since $P \subseteq P'$ and $P' \cap S = \emptyset$, we have $P' \cap R = P$.

DEFINITION. The sequence of elements $x_1, \ldots, x_n$ in the Noetherian ring R is a strong asymptotic sequence if $(x_1, \ldots, x_n) \neq R$ and for each $i = 1, \ldots, n$ $x_i \notin \cup \{P \in A_*((x_1, \ldots, x_{i-1}))$.

Remarks: (a) x_1 is a strong asymptotic sequence if and only if x_1 is regular ($A_*(0) = \operatorname{Ass} R$).

(b) If $(x_1, \ldots, x_n) \cap S = \emptyset$ with S multiplicatively closed, $x_1, \ldots, x_n$ a strong asymptotic sequence in R implies it is one in R_S.

(c) $x_1, \ldots, x_n$ is a strong asymptotic sequence in a local ring if and only if it is one modulo any $Q \in \operatorname{Ass} R$.

(d) R-sequences are strong asymptotic sequences (using $A_*(I) \subseteq A^*(I)$ and the fact that $A^*(x_1, \ldots, x_{n-1}) = \operatorname{Ass} R/(x_1, \ldots, x_{n-1})$ if $x_1, \ldots, x_{n-1}$ is an R-sequence).

(e) Strong asymptotic sequences are asymptotic sequences (roughly speaking, to be asymptotic, a sequence must behave well modulo any minimal prime in $(R_P)^*$ for any prime containing it. However strong asymptotic sequences behave well modulo any prime divisor of zero).

(f) If $R \subseteq T$ is a faithfully flat extension of Noetherian rings, $x_1, \ldots, x_n$ in R is a strong asymptotic sequence in R if and only if it is one in T. (Proposition 11.35.)

(g) Repeating, we do not know if this definition of strong asymptotic sequence coincides with that at the end of Chapter X. It does if $n = 1$ or 2.

PROPOSITION 11.36. Let I be an ideal in a Noetherian ring R and let $x_1, \ldots, x_n$ be a strong asymptotic sequence, maximal with respect to coming from I. Then $n = \min\{\text{depth } q^* \mid q^* \in \text{Ass}(R_P)^* \text{ for } I \subseteq P \in \text{Spec } R\} = \min\{\text{depth } q^* \mid q^* \in \text{Ass}(R_P)^* \text{ for } P \in A_*(I)\}$.

Proof: As the images of $x_1, \ldots, x_n$ form a strong asymptotic sequence in $(R_P)^*/q^*$ for any $P \supseteq I$ and $q^* \in \text{Ass}(R_P)^*$, we easily see height $(x_1, \ldots, x_n)(R_P)^* + q^*/q^* = n$. Thus n is equal to or less than the minimum of the first set above, which in turn is clearly equal to or less than the minimum of the second set. Finally, the maximality of our sequence shows there is a $P \in A_*(x_1, \ldots, x_n)$ with $I \subseteq P$. By Proposition 11.31, there is a $q^* \in \text{Ass}(R_P)^*$ with $(P_P)^*$ minimal over $(x_1, \ldots, x_n)(R_P)^* + q^*$. Clearly depth $q^* = \text{height } (P_P)^*/q^* = n$. However $(P_P)^*$ is obviously also minimal over $I(R_P)^* + q^*$ so that by Proposition 11.31, $P \in A_*(I)$. Thus the minimum of the second set is equal to or less than n.

Proposition 11.36 shows the following definition is unambiguous.

DEFINITION. For I an ideal in a Noetherian ring, let the strong asymptotic grade of I, $\text{gr}_* I$, be the length of a strong asymptotic sequence maximal with respect to coming from I.

We now state some results without proof. They are not hard, using the forgoing machinery.

PROPOSITION 11.37. $\text{gr } I \leq \text{gr}_* I \leq \text{gr}^* I$.

PROPOSITION 11.38. If I is an ideal in a local ring (R,M), then $\text{gr}_* I = \text{gr}_* IR^* = \min\{\text{height } IR^* + q^*/q^* \mid q^* \in \text{Ass } R^*\}$.

PROPOSITION 11.39. The following are equivalent for a Noetherian ring R.

i) $gr_* I = \text{height } I$ for all ideals I.

ii) $gr_* M = \text{height } M$ for all maximal ideals M.

iii) R is locally unmixed (i.e. depth $q^* = \dim R_P$ for every $P \in \operatorname{Spec} R$, $q^* \in \operatorname{Ass}(R_P)^*$).

If (R,M) is a local ring and $P \in \operatorname{Spec} R$, with $S = R - P$, then $(R^*)_S$ and $(R_S)^*$ do not appear to be closely related. Therefore the next result probably requires strong asymptotic sequences, or something similar. It can be proved by showing both numbers involved equal $gr_* P_S$.

PROPOSITION 11.40. Let $P \in \operatorname{Spec} R$, with R local. Then $n = m$ where $n = \min\{\text{depth } q^* \mid q^* \in \operatorname{Ass}(R_P)^*\}$ and $m = \min\{\text{height } P^*/Q^* \mid Q^* \subseteq P^*$ are primes in R^*, $Q^* \in \operatorname{Ass} R^*$, and $P^* \cap R = P\}$.

The analogue of Proposition 11.40, which talks about minimal primes rather than associated primes of zero is also true. Its proof uses $gr^* P_S$.

APPENDIX: Chain Conditions

This brief appendix states facts referred to in the preceding text, with enough references that the interested reader may pursue the matter. One new result is presented here.

DEFINITIONS. (All rings will be Noetherian)

i) A chain of primes $P_0 \subset P_1 \subset \ldots \subset P_n$ is <u>saturated</u> if height $P_{i+1}/P_i = 1$ for $i = 0, 1, \ldots, n-1$.

ii) A saturated chain $P_0 \subset \ldots \subset P_n$ is a <u>maximal chain</u> if P_0 and P_n are respectively minimal and maximal primes.

iii) The ring R is <u>catenary</u> if any two saturated chains of primes with common end points have common length.

iv) The local ring (R,M) is <u>quasi-unmixed</u> if every minimal prime in the completion R^* has depth equal to dim R.

v) The domain R satisfies the <u>Altitude Formula</u> if for any finitely generated extension domain of R, T, and for any $Q \in \operatorname{spec} T$ with $P = Q \cap R$, we have height $P + \operatorname{TrD}(T/R) =$ height $Q + \operatorname{TrD}((T/Q)/(R/P))$. Here $\operatorname{TrD}(B/A)$ is the transcendence degree of the domain B over the subdomain A.

Remark: It is well known that the Altitude Inequality holds for any Noetherian domain. That is, with R, T, P and Q as above, height $P + \operatorname{TrD}(T/R) \geq$ height $Q + \operatorname{TrD}((T/Q)/(R/P))$ [ZS].

It was long known that affine rings are catenary. As a consequence of his structure theorem for complete local rings, Cohen showed that complete local domains are catenary. Nagata's celebrated example [N , Example 2, pp. 203-205] gives a Noetherian domain R which is catenary but for which R[X], X an indeterminate, is not catenary. The following pair of results are fundamental (see [R1] or [MD]).

THEOREM A1. Let R be a Noetherian ring, and let X be an indeterminate. The following are equivalent.

i) Every finitely generated extension of R is catenary.

ii) for each maximal ideal M of R, $R[X]_{(M,X)}$ is catenary.

iii) R is locally quasi-unmixed.

iv) R_M is quasi-unmixed for each maximal ideal M of R.

THEOREM A2. Let R be a Noetherian domain. Then R satisfies the Altitude Formula if and only if R is locally quasi-unmixed.

We note that being quasi-unmixed is quite stable.

THEOREM A3. If the local ring (R,M) is quasi-unmixed, so is every finitely generated extension. If I is an ideal of R and if all minimal primes of I have the same depth, then R/I is quasi-unmixed.

Theorem A1 ((ii) $\Leftrightarrow$ (iv)) has been generalized (see [MD] or [RM]).

THEOREM A4. Let (R,M) be a local domain and let X be an indeterminate. Let $N \in \operatorname{spec} R[X]$ with $N \cap R = M$ but $N \neq MR[X]$. Then the following three sets of integers are all equal.

a) $\{n \mid$ there is a maximal chain of prime ideals of length n in $R[X]_N\}$

b) $\{n \mid$ there is a maximal chain of prime ideals of length n-1 in some integral extension domain of $R\}$.

c) $\{n \mid$ there is a minimal prime of depth n-1 in $R^*\}$.

If $R \subseteq T$ is an integral extension with (R,M) local, and if T contains a height 1 maximal then by going up to $\overline{T}$ and then going down to $\overline{R}$, we see that $\overline{R}$ contains a height 1 maximal. Thus Proposition 3.19 ((i) $\Leftrightarrow$ (iv)) is a special case of Theorem A4.

The next result is referred to in Chapter 5 and Chapter 10. As it does not appear in the literature, we prove it.

THEOREM A5. Let P be a prime in a local ring (R,M). Suppose that $(R/P)^*$ has a depth n minimal prime and $(R_P)^*$ has a depth m minimal prime. Then R^* has a depth $n+m$ minimal prime.

Proof: By Theorem A4, there is a maximal chain of length $n+1$ in $(R/P)[X]$ localized at $(M/P,X)$. Under the natural map $R[X] \to (R/P)[X]$, suppose the inverse image of that chain is $PR[X] \subset P_1 \subset \ldots \subset P_n \subset (M,X)$. Now this chain is saturated, and by [HM, Corollary 1.5] we may assume that $P_1 \cap R = P$ (but $P_1 \neq PR[X]$).

Considering $R \to R_P \to (R_P)^*$, we let q^* be a depth m minimal prime of $(R_P)^*$ and q be its inverse image in R. Thus $q^*/q(R_P)^*$ is a depth m minimal prime in $(R_P/qR_P)^* = ((R/q)_{P/q})^*$. By Theorem A4, applied to $(R/q)_{P/q}$ with $N = P_1R_P[X]/qR_P[X]$, we easily find a saturated chain of primes of length $m+1$ having the form $qR[X] \subset Q_1 \subset \ldots \subset Q_m \subset P_1$.

Clearly $qR[X] \subset Q_1 \subset \ldots \subset Q_m \subset P_1 \subset P_2 \subset \ldots \subset P_n \subset (M,X)$ is a maximal chain of length $m+n+1$. Applying Theorem A4 to $R[X]_{(M,X)}$, we see that R^* contains a depth $m+n$ minimal prime.

REFERENCES

[B1] M. Brodmann, "Asymptotic stability of Ass (M/I^nM)", Proc. Am. Math. Soc., 74(1979), 16-18.

[B2] ________, "Asymptotic nature of analytic spreads", Math. Proc. Camb. Phil. Soc., 86(1979), 35-39.

[B3] ________, "Uber de Minimal Dimension der Assozuerten Primeideale der Kompletion eines Lokalen Integritatsbereiches", Comment. Math. Helv., 50(1975), 219-232.

[B4] ________, "Piecewise catenarian and going between rings", Pacific J. Math., 86(1980), 415-419.

[BR] M. Brodmann and C. Rotthaus, "Local domains with bad sets of formal prime divisors", J. Algebra, 75(1982), 386-394.

[Bu] L. Burch, "Codimension and analytic spread", Proc. Camb. Phil. Soc., 72(1972), 369-373.

[D] E. Davis, "Ideals of the principal class, R-sequences, and a certain monoidal transformation", Pacific Math. J., 20(1967), 197-205.

[E] E.G. Evans, "A generalization of Zariski's Main Theorem", Proc. Am. Math. Soc., 26(1970), 45-48.

[ES] P. Eakin and A. Sathaye, "Prestable ideals", J. Algebra, 41(1976), 439-454.

[FR] D. Ferrand and M. Raynaud, "Fibres formelles d'un anneau local Noetherian", Ann. Sci. Ecole Norm Sup., 3(1970), 295-311.

[H1] R. Heitmann, "Prime ideal posets in Noetherian rings", Rocky Mountain J. Math., 7(1977), 667-673.

[H2] ________, "A non-catenary normal local domain", Rocky Mountain J. Math., 12(1982), 145-148.

[HM] E.G. Houston and S. McAdam, "Rank in Noetherian rings", J. Algebra, 37(1975), 64-73.

[K1] I. Kaplansky, Commutative Rings, University of Chicago Press, 1974.

[K2] ________, "Adjacent prime ideals", J. Algebra, 20(1972), 94-97.

[Kz1] D. Katz, "Asymptotic primes and applications", Ph.D. Dissertation, University of Texas at Austin, 1982.

[Kz2] ________, "A note on asymptotic prime sequences", Proc. Am. Math. Soc., (to appear).

[Kr] W. Krull, "Zum Dimensionsbegriff der Idealtheorie" (Beitragle zur Arithmetik Kommutativer Integritatsbereiche, III), Math. Z., 42(1937), 745-766.

[M1] S. McAdam, "1-going down", J. London Math. Soc., 8(1974), 674-680.

[M2] ________, "Saturated chains in Noetherian rings", Indiana Univ. Math. J., 23(1974), 719-728.

[M3] __________, "Asymptotic prime divisors and going down", Pacific J. Math., 91(1980), 179-186.

[M4] __________, "Asymptotic prime divisors and analytic spreads", Proc. Am. Math. Soc., 80(1980), 555-559.

[MD] S. McAdam and E. Davis, "Prime divisors and saturated chains", Indiana Univ. Math. J., 26(1977), 653-662.

[ME] S. McAdam and P. Eakin, "The asymptotic ass", J. Algebra, 61(1979), 71-81.

[Ma] J. Matijevic, "Maximal ideal transforms of Noetherian rings", Proc. Am. Math. Soc., 54(1976), 49-51.

[Mt] H. Matsamura, Commutative Algebra, Benjamin, 1970.

[N] M. Nagata, Local Rings, Interscience, 1962.

[N_i] J. Nishimura, "On ideal transforms of Noetherian rings II", J. Math. Kyoto Univ., 20(1980), 149-154.

[O] T. Ogoma, "Non-catenary pseudo-geometric normal rings", Japan J. Math., 6(1980), 147-163.

[R1] L.J. Ratliff, Jr., "On quasi-unmixed local domains, the altitude formula, and the chain condition for prime ideals (I)", Amer. J. Math., 91(1969), 508-528.

[R2] __________, "On quasi-unmixed local domains, the altitude formula, and the chain condition for prime ideals (II)", Amer. J. Math., 92(1970), 99-144.

[R3] __________, "On prime divisors of I^n, n large", Michigan Math. J., 23(1976), 337-352.

[R4] __________, "Two theorems on the prime divisors of zero in completions of local domains", Pacific J. Math., 81(1979), 537-545.

[R5] __________, "A(X) and GB-Noetherian rings", Rocky Mountain J. Math., 9(1979), 337-353.

[R6] __________, "Integrally closed ideals and asymptotic prime divisors", Pacific J. Math., 91(1980), 445-456.

[R7] __________, "Note on asymptotic prime divisors, analytic spreads and the altitude formula", Proc. Am. Math. Soc., 82(1981), 1-6.

[R8] __________, "On asymptotic prime divisors", Pacific J. Math., (to appear).

[R9] __________, "Asymptotic sequences", (manuscript).

[RM] L.J. Ratliff, Jr. and S. McAdam, "Maximal chains of prime ideals in integral extension domains I", Trans. Am. Math. Soc., 224(1976), 103-116.

[RR] L.J. Ratliff, Jr. and D. Rush, "Two notes on reductions of ideals", Indiana Univ. Math. J., 27(1978), 929-934.

[Rs1] D. Rees, "Valuations associated with ideals II", J. London Math. Soc., 36(1956), 221-228.

[Rs2] __________, "Rings associated with ideals and analytic spreads", Math. Proc. Camb. Phil. Soc., 89(1981), 423-432.

[S1] J. Sally, "Bounds on generators of Cohen-Macaulay ideals", Pacific Math. J., 63(1976), 517-520.

[S2] ________, "A note on integral closure", (manuscript).

[Sc] P. Schenzel, "Independent elements, unmixedness theorems, and asymptotic prime divisors", (manuscript).

[Sm] P. Samuel, "Some asymptotic properties of powers of ideals", Annals of Math., 56(1952), 11-21.

[SO] M. Sakuma and H. Okuyama, "On a criterion for analytically unramification of a local ring", J. Gakugel, Tokushima Univ., 15(1966), 36-38.

[W] K. Whittington, "Prime divisors and the altitude formula", Ph.D. Dissertation, University of Texas at Austin, 1980.

[ZS] O. Zariski and P. Samuel, Commutative Algebra, vol. II, D. Van Nostrand, 1980.

LIST OF NOTATION

(Page numbers indicate where more information can be found.)

		Page
$A(I,n)$	$\mathrm{Ass}(R/I^n)$	3
$A^*(I)$	limit of $A(I,n)$, $n = 1,2,\ldots$,	
$\bar{A}(I,n)$	$\mathrm{Ass}(R/\overline{I^n})$	
$\bar{A}^*(I)$	limit of $\bar{A}(I,n)$, $n = 1,2,3,\ldots$,	12
$B(I,n)$	$\mathrm{Ass}(I^{n-1}/I)$	
$B^*(I)$	limit of $B(I,n)$, $n = 1,2,\ldots$,	3
$\bar{B}(I,n)$	$\mathrm{Ass}(\overline{I^{n-1}}/\overline{I^n})$	
$\bar{B}^*(I)$	limit of $\bar{B}(I,n)$, $n = 1,2,3,\ldots$,	100
$gr^*(I)$	the asymptotic grade of I	35
$\bar{I}$	the integral closure of the ideal I	3
$I \sim J$	projective equivalence	53
$\ell(I)$	the analytic spread of I	26
$Q(R)$	the total quotient ring of R	
$\bar{R}$	the integral closure of R in $Q(R)$	
R^*	the M-adic completion of the local ring (R,M)	
$R^{[1]}$	$\{y \in Q(R) \mid (R:y) \not\subseteq P$ whenever $z(P) = 1\}$	77
$R^{\langle 1 \rangle}$	$\{y \in Q(R) \mid (R:y) \not\subseteq P$ whenever $\mathrm{Ass}((R_P)^*)$ contains a depth 1 prime$\}$	87
$T(I)$	the ideal transform of I	76
$V_I(x)$	n if $x \in I^n - I^{n+1}$; ∞ if $x \in \cap I^n$, $n = 1,2,3,\ldots$	
$\bar{V}_I(x)$	$\lim_{n\to\infty} V_I(x^n)/n$	39
$z(P)$	$\min\{\text{depth } q^* \mid q^* \text{ is a minimal prime of } (R_P)^*\}$	
$\subseteq$, $\subset$	inclusion , proper inclusion	

INDEX

index (continued)